Formal App[...] Computing
and Information Technology

Also in this series:

Proof in VDM: a Practitioner's Guide
J.C. Bicarregui, J.S. Fitzgerald, P.A. Lindsay, R. Moore
and B. Ritchie
ISBN 3-540-19813-X

On the Refinement Calculus
C. Morgan and T. Vickers (eds.)
ISBN 3-540-19809-1

Systems, Models and Measures

by Agnes Kaposi and Margaret Myers

Springer-Verlag
London   Berlin   Heidelberg   New York
Paris   Tokyo   Hong Kong
Barcelona   Budapest

Agnes Kaposi
Kaposi Associates
3 St Edwards Close
London NW11 7NA, UK

Margaret Myers
33 Sneath Avenue
London NW11 9AJ, UK

Series Editor

Steve A. Schuman, BSc, DEA, CEng
Department of Mathematical and Computing Sciences
University of Surrey, Guildford, Surrey GU2 5XH, UK

ISBN 3-540-19753-2 Springer-Verlag Berlin Heidelberg New York
ISBN 0-387-19753-2 Springer-Verlag New York Berlin Heidelberg

British Library Cataloguing in Publication Data
A catalogue record for this book is available from the British Library

Library of Congress in Publication Data
A catalog record for this book is available from the Library of Congress

Typesetting: Camera ready by authors
Printed and bound by Antony Rowe Ltd, Bumper's Farm,
Chippenham, Wiltshire
34/3830-543210 Printed on acid-free paper

To our husbands for their patience

Preface

This book has been written in the interest of quality. It is motivated by the realization that quality is implicit in all purposeful human activity. The quality of the questions we pose, the options we generate, and the decisions we make on the way to a solution, determine the quality of the outcome. Quality distinguishes between the success or failure of our endeavours, and the utility or futility of our contributions.

The notion of quality is incompletely understood, and one must acknowledge that some of its aspects are undefinable. Yet we all appreciate the quality of the artefacts we use, and we are conscious of the quality of the service we receive (or the lack of it). We value the quality of debates with our colleagues and of the companionship of friends, and we are aware of the need to protect the quality of our environment. It is one of the obligations of science to offer us deeper insight into a notion of such global significance, and it is the entire purpose of technology to give us the means to guard and enhance quality.

The approach to quality leads through an understanding of *systems*, *models* and *measurement*. The concept of 'system' is necessary for making sense of observations in a world of limitless variety, and to formulate ideas in terms of powerful general principles. Measurement is the means of recording observations about each particular situation. Models are formal links between the specifics of the problem in hand, and the general organizing principles needed to assure a good solution. Systems concepts, modelling methods and practical measures are required for taking control of problematic situations: perceiving the need, setting the task, formulating possible solutions, making expert choices, and passing informed judgements on the quality of the results.

Metrology has been the foundation of the impressive development of the natural sciences and their technologies. Lack of such foundations in other fields, such as the social sciences and software, is a major reason for their lesser scientific development, and this now hinders progress in the classical sciences which in turn need the new disciplines for further advance. The applicability of science relies on effective coordination of the classical and the new: real-life problems are 'systems' problems which transcend discipline boundaries.

The answer does not lie in developing distinct 'quantitative methods' and inventing a new cluster of 'metrics' for the young disciplines. Not only is this an inefficient re-invention of some of the well-crafted wheels of science: the insights so gained are of narrow scope and transient validity. Measures and methodologies, like the problems themselves, must transcend domain boundaries and must be suited to many different fields of application. For this reason the book advocates using systems concepts and formal models, and the grounding of the design of new measures in the extant knowledge base of classical metrology, in accord with the scientific method.

The book proclaims that a measurement system must be *model-based*. The models put in context the (objective) facts of the situation, and assist in secure reasoning about them. They can also make explicit the problem solver's (subjective) goals and views. Thus, a model-based measurement scheme aids discrimination between fact and value judgment, and guards the quality of the solution.

Exploiting the power of systems concepts over the universal domain, a model-based measurement scheme is adaptable to any application area, and is suited to inter-disciplinary use. It offers potential for the development of a metrological system whose scope extends to the new and the classical sciences, and allows the measurement of quality, or, more precisely, those aspects of quality which can be captured in well-defined concepts. Models can treat the target system and the environment symmetrically, thus allowing the characterization by measurement of either from the viewpoint of the other, and dealing even-handedly with them both.

The claim of systematic generalization without trivialization needs validation. Validation of the feasibility of model-based measurement can be achieved through examples, experiments, or experience in practical applications. The book draws on all three. Practical model-based measurement examples are quoted from a variety of fields, mostly relying on the comprehensively validated models and measures of natural science, but, in order to test the feasibility of model-based measurement in some new domain where measurement experience is still scarce, the book provides its own extended case study demonstration.

The application domain selected for the case study is that of formal specifications – an area of particular importance from the viewpoint of quality. The case study demonstrates the possibility of modelling what is expressed in the formal specification, and measuring what is modelled. The aim of the study is to validate the procedure of model-based measurement. The models do not attempt to provide a definitive characterization, nor do they advocate qualitative judgement of specifications, but they capture key attributes of the specifications which are relevant to their 'complexity', such as their data structure, organization, and the dimensions of their data domain. To illustrate the wide applicability of the organizing principles, the case study links together several specification media within a multiple-language measurement scheme. The possibilities of the case study are far from being fully explored in the book; the experiment could (and should) be continued, and the measurement scheme extended. Such further work would allow the measures to reveal emergent properties of specifications expressed in declarative languages, which could then lead to new insights into the quality of specifications, designs, and the formal languages themselves.

Model-based measurement is also put to the independent test of practical application in solving problems in many diverse fields. In recent years, model-based measurement has been exercised in various applications, including software development, project management, industrial strategy formation, and the development of technical (as opposed to just administrative) quality systems for various branches of industry. In each case, the use of model-based measurement arose naturally, and has indeed led to deeper understanding of the issues at hand, clearer specifications, deliberate solution strategies, and greater confidence in the results.

Using the book

Since it builds practice on first principles, the book should hold interest for many types of readers: researchers and practitioners, technologists and managers, students and lecturers, system procurers and suppliers, standard setters and standard enforcers. Since its scope extends over any application area and discipline domain, the book aids all problem solvers, but it would offer particular benefits for:

- scientists and engineers who work on, or beyond, the boundaries of the natural sciences, needing the extension of classical metrology;
- scientists and engineers who work in new and developing sciences and technologies, where sound measurement foundations are yet to be established;
- practitioners whose job is to specify, design, implement and operate complex systems, or manage the processes associated with such systems;
- managers concerned with making correct, effective and fair decisions about products, services, finance, personnel, other key resources, and the environment.

To such readers, the book offers help in conceptualization, communication, characterization, and decision making.

Conceptualization and communication

The once-renowned Professor Joad, in his famous catch-phrase "It all depends on what you mean by . . .", popularized the truth that words are not always tightly coupled to meaning. Many shallow ideas hide behind impressive jargon and attractive slogans. Many arguments have failed to reach a conclusion, or have led to agreements which later proved to have been based on misunderstanding, either because of the vagueness of the concept or because of the ambiguity of the terminology. The difficulties are not confined to lay people grappling with bewildering technical notions: problems of communication arise within inter-disciplinary professional teams even when, within each individual discipline, the concepts are clear and the terminology consistent. (We believe that whenever disagreement arose in course of the compilation of this book, the cause lay in the authors' differing backgrounds, one being in engineering and the organizing sciences of systems, and the other in experimental science and computing. The resolution of such arguments are embodied in the material of the book.)

Systems science strengthens the hand of the problem solver by assuring that many sophisticated ideas can be captured in a few, well-chosen, clearly-defined concepts. This is why we have taken care to identify and define the key notions of system, model and measure, and others, such as artefact, product and process; attribute and property; black-box behaviour and structure; specification and design. Most of these concepts are used in technological practice widely, but often casually. We have sought to provide definitions of adequate quality, avoiding circularity, and have attempted to use the terminology consistently, reinforcing definitions by mathematical representations where necessary. The power and elegance of these general concepts, and the frugality of their number, helps conceptualization of problems, and facilitates communication between problem owners and problem solvers, whatever their background.

Characterization and decision making

Characterization of an entity calls for objective observation. Rational decision-making calls for explicit declaration of subjective judgement based upon objective observation. To the discerning observer, the scientist, the technologist and the manager, the book presents a generic measurement scheme which imposes order over observations, ties attributes to their referents, and systematizes objective and subjective measures. Such a scheme makes sophisticated attributes of complex entities measurable, and reassures the user as to the validity of the measures.

Measurement of quality

The standard definition of quality of an 'entity' is said to be: "the totality of features and characteristics . . . that bear on its ability to satisfy stated or implied needs". The definition indicates a practical approach to the notion of quality which excludes consideration of its undefinable 'transcendental' aspects. To assure quality (in the sense of demonstrating it beyond reasonable doubt) means that the user compares the product or service on offer with the needs, and decides on the consistency of one with the other. The supplier often has a hard time persuading the customer into a favourable judgement on quality, even when the needs are stated clearly in the specification. How very much harder is the task (and what rich grounds there are for disputes and disagreements!) when the needs are not stable but may evolve in time, and are not stated, but merely "implied"!

This is just one instance where model-based measurement comes into its own. Systems notions and models aid the user to conceptualize the 'features and characteristics' of the entity, define them in variables, and communicate these to the supplier explicitly, rather than leaving them to remain 'implied'. In cooperation with the user, the supplier is able to express specifications formally (and may even animate them!), identifying key attributes, and defining them in measurable terms. Together, the user and the supplier can explicitly set the objective measures which would satisfy the needs, and articulate subjective measures on which quality will be ultimately judged. Instead of vagueness, potential strife, and punitive quality control, model-based measurement provides the grounds for a mature, constructive approach to quality.

Adopting model-based measurement in industry

Upheavals are the last things needed by any functioning organization, and yet systematic model-based measurement would be a radical departure from current technological, managerial and quality practice. How then can the methodology of model-based measurement be implemented in companies, businesses and the like? We suggest progressing naturally through stages:

(1) Introduction of the key concepts of systems and model-based measurement by education throughout the organization, as part of the quality policy.

(2) At operational levels, a critical inspection of information collection and decision-making procedures, and an assessment of how these correlate with model-based measurement.

(3) The identification of a suitable referent problem by individuals; the proposal of a modelling syntax and measurement scheme; the critical assessment of short- and long-term costs and benefits by the individual.

(4) Discussion among individuals of experiences and observations.

(5) Reinforcement of education and information exchange, and a widening of the procedure to larger groups of participants.

(6) Evaluation of the perceived costs/benefits at various organizational levels.

(7) Rationalization of old practices and development of new practices, all based on the evaluation.

Future developments

The ideas and experiences described in this book may be explored and utilized in several ways:

- Formalizing the theoretical foundations of model-based measurement.
- Investigating the possibility of extending the classical metrological system to new domains.
- Extending the case study in two ways: involving more formal languages in the multiple-language measurement strategy, evaluating the measures and developing them in the light of wider experience with their use.
- Gathering broader, more detailed, experience by devising model-based measurement schemes in various application-specific and industry-specific fields.
- Using model-based measurement as an integral part of quality systems within companies and other organizations.
- Using model-based measurement as a means of reviewing and enhancing the quality of standards.

Those who work within the framework of model-based measurement will come to find that they use the same concepts, share the same language, employ a consistent methodology, and derive the same benefits from being part of a common culture.

London   Agnes Kaposi
August 1993   Margaret Myers

Acknowledgements

The authors are grateful to John Kaposi for his professional help with the structure, content, and form of the book. They also thank Professors B. Cohen and S. Schuman for their constructive comments and suggestions on early drafts.

The authors thank their many friends and colleagues at BT Martlesham for their stimulating questions and their support for some of the work, and in particular, Alan Stoddart and Charles Jackson.

Contents

Part 2

PART 1

1 INTRODUCTION

The context

The first half of this century was marked by the development of the natural sciences and their technologies. Measurement is the foundation of all scientific development, and rapid progress in these fields was facilitated by a comprehensive, theoretically sound, and universally accepted system of measurement.

The second half of the century proves to be a period of technological division and specialization. New sciences emerge and new technologies are created, such as computer science and the social sciences, artificial intelligence, management, biotechnology and software engineering. These new disciplines formulate important new concepts, grow their own roots, introduce their own terminology, and construct novel types of systems whose attributes are beyond the scope of classical science and measurement. While the new sciences and technologies bring about unprecedented progress, their separate development has the adverse effect of dividing the scientific community. After an initial period of rapid development, the cleavage between the mature and novel sciences and technologies is proving a hindrance to them both. The harsh problems of today have no regard for discipline boundaries: their effective solution demands coherent scientific foundations, and close co-ordination between the mature and novel technologies.

We now enter the systems era: a period of growing recognition of the need to bridge the gap between the sciences and technologies of the classical fields and those of the new disciplines, and create for the comon foundations. This book aims to contribute to this end.

The book

The message of the book is three-fold. The first is, by now, almost a cliché: measurement is essential for the identification, formulation and solution of problems. The second is that models are the prerequisite of measurement. Emphasizing this, we have coined the phrase 'model-based measurement'.

The third aspect of the message is that, using the unifying notion of systems, one may devise a general scheme for model-based measurement, equally suited to the traditional sciences and the new technologies. Otherwise there can be no effective co-ordination, and no possibility of building the new sciences and technologies on the foundations of the old.

Problem solving by common sense or science?

Industry and business, scientific and professional practice, work and leisure, all give rise to problems, constantly calling for decisions. Most problems of everyday life can be solved promptly and informally, on the basis of experience, common sense and intuition. But while these are indispensable in every situation, in themselves they are an inadequate basis for solving the managerial, technological and scientific problems of a modern society. Circumstances change so fast that yesterday's experience and today's practices may be obsolete by tomorrow; the requirements may be so diffuse that the nature of the problem may not be intuitively understood; the solutions are seldom derivable from the problem solver's own experience alone, but instead call for the combined skill and concerted effort of experts in many fields; the margin between good and bad decisions may be too narrow, and the consequences too severe, for the decision to be trusted to common sense. In the fiercely competitive scene of industry and business, managers have become increasingly accountable, called upon to make explicit the grounds on which value judgements are made. Practical as such problems are, strong scientific foundations are needed for their effective solution.

Why do we need science and technology?

Science is the accumulation and distillation of experience. People observe the world, record their observations, make comparisons, and draw inferences. As experience widens and understanding grows, patterns are discerned, hypotheses are formed [1], and predictions are made about trends and future events. If these are independently validated then theories follow and from these new truths can be derived. This is how a science is born.

Technology is the purposeful use of science: the bridge between science and its enlightened application. In the widest sense, as defined by Bunge [2], technology embraces all domains of practical endeavour. Bunge points to four branches of technology, according to their main scientific foundations: physical technologies (such as mechanical engineering and electronics), biological technologies (such as medicine and pharmacy), social technologies (such as law, education and the social services), and thought technologies (such as computing and artificial intelligence).

Science, technology and application advance by mutual reinforcement. The success of modern technological societies rests on scientific foundations; only on the grounds of good science can we build sound technologies, to be employed for the public good. Technologies facilitate the perception, recording and analysis of needs, provide skill and tools of observation, stimulate ingenuity of design, afford integrity of validation, and promote efficiency of implementation of solutions. Society advances through the interaction and mutual reinforcement of science, technology and application.

Why do we need measurement?

The reliance of the natural sciences and their technologies on measurement is well understood. Scientific results and technological achievements can only be judged on the basis of evidence, and convincing evidence can only be provided by measurement. Kyburg [3] introduces his excellent book on measurement by the sentence: "Measurement is so fundamental to the physical sciences and to engineering that it is difficult to know what we would do without it." Krantz [4] talks of the "crucial concern with measurement" in the physical and the behavioural sciences. Katzner [5] emphasizes the role of (non-numerical) measurement in the solution of managerial and political problems. Roberts [6] broadens the scope even further: he quotes such global problems as pollution, energy utilization, and fair allocation of the world's resources, noting that "the ability to measure is critical" to the tackling of all such problems.

One may sum up the teaching of these and other distinguished authors: measurement is the key to all disciplines of science and technology, and the maturity of a discipline is marked by the extent to which it is supported by a sound and comprehensive system of measures, measurement standards, measurement tools and measuring procedures.

This case for the importance of measurement is compelling, but it gives only part of the answer to the question "why measure?". The need to measure is universal: it is not confined to scientists, technologists and the professional problem solvers of sophisticated technological societies, but essential for the intuitive problem solver, and in the conduct of the business of everyday life of all societies. We measure because only through measurement can we know the world around us, understand and formulate problems, generate possible solutions, and evaluate the suitability of the solution. Measurement is more than the recording of facts by collection of data: it is also the basis of expressing values and forming judgements. Measurement is the instrument by which to assure the quality of products and services, and the fairness of business and trade.

What *is* and what *is not* measurement?

Measurement is both science and technology. It is a systematizing science, covering all sciences, and an essential part of all technologies. Up to the 1950s, measurement was regarded, and taught, as a branch of technology in its own right, but it seldom has such prominence in the modern curriculum. It may be considered a key subject again in the future, in view of the mounting measurement problems of complex cross-disciplinary systems.

The purpose of measurement is to provide a valid, trustworthy, traceable representation of some chosen target entity, called the 'referent', whose selected attributes are of interest. The referent may be an existing object or phenomenon, or else some future system, to be described by a specification, refined by design, and realized by implementation.

The popular notion of measurement is that it gives a quantitative representation. It is true that measurement usually involves the use of numbers, but data collection and casual assignment of numbers to things do not constitute measurement. Since numerical descriptions inspire confidence, quantification is frequently used to lend bogus authority to decisions, and the results are mistakenly referred to as 'measurements', even when the proper scientific foundations of measurement are lacking. As all powerful tools, numerical representations can be misused, such that the numbers (which are sometimes called 'metrics') give an unreliable or misleading impression, rather than a meaningful, traceable, and representative description of the referent. The 'metrics' may make little sense, or give a false image of the problem, and misrepresent the intended solution. The result may be a failure of the project: the error may only come to light after implementing a tentative solution which then fails validation, or, even worse, after the invalid solution fails in practice. By then, time and resources have been wasted, opportunity lost, and the end user may have been put at risk.

While some numerical descriptions do not constitute measurement, some of the most important measures of science, technology and everyday life are non-numerical, employing symbols other than numbers to describe the referent. Alphabetic characters, colours, pictures, sounds, and other symbols can effectively signify measures such as designation of blood types, grading of specifications, alarm conditions and marks of quality approval. The new set of British bank notes bear numerals too small for the partially sighted, the sizes of notes of different value are similar, and the ambiguous colours (devised to confound the forger) are not distinctive. Instead, it is claimed, the value of the note is most effectively designated by printing on it a distinct shape: square for £20, diamond for £10, circle for £5, etc.

What to measure?

The target of measurement may be any class of entity: a physical phenomenon, a material object, a software product or a complete hardware/software system, a process relating to the creation, operation and use of the system, or the environment in which the system operates. The measured entity may also be a service, or a service provider, such as a person or organization responsible for software maintenance.

What is modelling, and what is model-based measurement?

Real-life problems are complicated, involving entities with many features, and parts which interact with each other and their environment in intricate ways. This means that many of the most important referents are complex, and it is hard to isolate their primary characteristics from their secondary features. Modelling replaces a complex system by a simpler 'model' system which is, nevertheless, representative. Modelling is our only known tool for coping with complexity.

A good model is not an arbitrary representation, but a systematic, purposeful simplification of its referent. Such a model focuses attention on selected attributes which hold the key to the solution of the given problem, and suppresses features which, for the given situation, are less crucial or irrelevant. A good model captures the key attributes in well-defined property variables, and preserves the inter-relations among attributes. The values assigned to the property variables by observation are the *measures* of the referent. This is the process which we call *model-based measurement*.

Model-based measurement proceeds in two steps:

(1) Given some referent entity of arbitrary complexity, its model is created: a simpler system which captures the selected attributes of the referent, and characterizes them by well-defined and explicitly related property variables.

(2) Given the model, a suitable formal system of numbers or other symbols is selected to describe possible values of each property variable. The elements of the set of property values are the measures of the model, and represent the referent.

Without the model, measures would be meaningless: they lack definition, context and interpretation; hence all proper measurement is *model-based*. It can be said that if only model-based measurement counts as proper measurement, then the qualification 'model-based' is redundant. Nevertheless, we advocate the use of the term, and maintain it throughout the book, for emphasis, and to make a clear distinction between the casual use of quantification, and soundly based, proper measurement.

What is a measure?

A measure is a value of a property variable of the model of the referent. If the model is a valid representation of the selected attributes of the referent, then the measure will be a valid representation of one of these attributes.

A measure does not always arise from objective observations: it may represent a desired attribute of a future referent under specification, or else it may prescribe a limit value which must be attained by some object such that it should meet standards. Nor is a measure always objective. It may be an orderly representation of subjective views or perceptions of members of the constituency of the referent system: the problem solver, the client, the prospective user, or the public.

What is a 'model-based measurement scheme'?

A single measure is seldom sufficient to describe the object of interest. Even if this is the case at the start, sooner or later attention will extend to other features, and some new measures will be needed to describe these.

The referent's attributes of interest may be open to *direct* observation; on the other hand, the value of the requisite property may have to be obtained by *indirect* measurement, where the measure is derived from other, directly accessible property values. The individual property measures may have to combine to form a coherent representation of the referent, where it is assured that they correspond to observations which are consistent, and not dispersed in time. Such a coherent representation is called *object-oriented measurement*. Measurement may also be directed at the formation of opinions about the object-oriented measures of the referent, so as to articulate value judgements and support decisions. This kind of explicit expression of subjective views is facilitated by *utility measurement* (7).

The direct, indirect, object-oriented and utility measures characterize the referent within a co-ordinated *measurement scheme*. Such a measurement scheme is a framework for characterizing all the attributes, models and measures of a referent, and one which also reconciles the apparent conflict between objectivity and value judgement. The models of the scheme make explicit the relationship which holds over the individual measures. A good measurement scheme is robust and dynamic: it can develop and expand by the modification and refinement of models and measures, and by admission of new measures and deletion of obsolete ones, as experience dictates and circumstances demand. The measurement scheme facilitates the evaluation and refinement of existing measures, and the development of new measures, while maintaining consistency.

The foundations of mature areas of science and technology incorporate their model-based measurement scheme, implicit in the proven theories of the discipline, and in the validated measures, international units, well-tried procedures and agreed standards of an established metrological system. Here the problem solver often knows in advance which specification features will play key roles in the decision-making process, can model the object in terms of these, can rely on well-defined models for deriving the indirect measures, and can assure that all (and only!) the relevant direct measures are collected. The competent practitioner can rely on the proven theories and validated measures of the discipline, which help to assure that the collection, interpretation and utilization of data is systematic.

The new and developing fields of technology and science have no such firm foundations. In some fields, such as software engineering and management, the theories are still being created, the key attributes of interest are still under debate, the use of measurement is not universal, at times it is hard to know what data to collect, what the data mean, and how the data items may relate to each other and to the project as a whole. There are attempts to propose and validate measures, but a consensus is yet to be reached as to what may constitute a sensible collection of measures. In such circumstances the design of a comprehensive measurement scheme would be premature, because some of the tools of obtaining direct measures and part of the basis of deriving indirect measures may be lacking, and the utility criteria may not yet be fully understood.

In such cases the principles of model-based measurement can be explicitly evoked to provide guidance to creating some early version of a suitable measurement scheme. Although a full complement of object-oriented properties is not yet defined and the definition of utility measures is not available, a valid model can usually be created to capture some of the referent's observable features. The model draws attention to direct measures which may be of interest in themselves – why else would they be preserved in the model? The direct measures can then form a 'knowledge base', and, from this, one may start to develop a more sophisticated version of the model-based measurement scheme. The scheme may then lead to the derivation of other measures, and experimentation with value measures. As the measurement scheme evolves, it may bring out emergent properties which lead to the discovery of new, previously unidentified attributes of the referent. In this way, the measurement scheme promotes understanding and underpins the maturing of the discipline.

Such inductive uses of model-based measurement are discussed in Part 1 of the book, and are illustrated in the case study of Part 2.

How to devise a model-based measurement scheme?

As we have noted before, the maturity of the classical disciplines of science and technology is derived from their implicit model-based measurement schemes. If the principles of model-based measurement are generally applicable, as this book claims, could one aid the maturing of some important new field of technology by devising for it a model-based measurement scheme? Or, putting it another way, can one produce evidence of the wide applicability of model-based measurement by devising a measurement scheme for some new field of technology?

As a case study demonstration, in this book a rudimentary model-based measurement scheme is devised and implemented, where the referent is the class of specifications expressed in formal languages. For the sake of breadth and efficiency, the class extends to more than one notation, by use of a 'multiple-language measurement strategy'. This means constructing a model-based measurement scheme for one of the formal languages – in this case Prolog – and using the scheme for capturing measurable attributes of specifi-cations written in several other languages, among them VDM, Z, and some semi-formal specification media, with the view to extending the scope to other specification languages later. The measures may be used for several purposes, such as characterizing individual specifications of a given referent, comparing specifications and designs, and comparing features of the specification languages themselves. The case study shows how the measures of the scheme may be read off directly from the models, or derived indirectly, by composition of measures, and how the measures of parts aggregate to form measures of complete specifications. The study also shows how the measurement has been automated, with the aid of software tools.

Note that the object of the demonstration is to validate the feasibility of composing and implementing a model-based measurement scheme, rather than to validate the measures themselves. Although the measures have been selected with care and have been used in two industrial investigations on specifications of several kinds, it is not claimed that collectively the measures of the scheme define some kind of industry standard which would constitute a comprehensive characterization of formal specifications. On the contrary: it is emphasized that, for the purposes of this demonstration of model-based measurement, the choice of measures is arbitrary, as long as the strictures of model-based measurement are observed. There are suggestions of alternative measures, and further developments of the measurement scheme.

The structure of the rest of the book

PART 1 presents the general foundations of model-based measurement, and describes a case study to validate the feasibility of devising a particular model-based measurement scheme for a new field.

Chapter 2 defines the ideas of 'system' and 'model', and builds up a four-part 'theory of models', moving from the relationship between referent and model through the description of the referent in its environment to the representation of the referent as a unity and as a structure of parts.

Chapter 3 is devoted to the discussion of the general principles and concepts of measurement, and the definition of the criteria of measurability. The notion of a 'model-based measurement scheme' is introduced: a general framework for characterizing the objectively observable attributes of systems of any kind by their models and measures, and representing explicitly subjective views and values relating to them. The general scheme of model-based measurement is developed step-by-step, as a four-layer structure.

Chapter 4 uses the notions of models and measures for defining and distinguishing product and process, and setting out the principles of specification and design of products and processes.

Chapter 5 shows how model-based measurement may be used to characterize the specification and design of products and processes, laying emphasis on a class of systems which have particular significance in the new technologies: finite state systems.

Chapter 6 presents the case study. It explains why the field of formal and semi-formal specifications has been chosen to demonstrate the development of a model-based measurement scheme. It describes the 'multiple-language strategy' of model-based measurement of specifications, and the criteria for choosing a common reference language for measurement. It summarizes the results and the conclusions which can be drawn from them.

PART 2 gives the details of the demonstration case study of Chapter 6.

Chapter 7 presents Prolog as the chosen reference language of the proposed multiple-language measurement strategy. It is shown that Prolog meets the criteria for a suitable reference language, but the choice is neither exclusive nor critical. This chapter outlines the structure and facilities of Prolog, identifying and isolating the 'Prolog logic text': the part of the language needed for its use in model-based measurement of specifications.

Chapter 8 and **Chapter 9** form a pair. They develop, respectively, models and measures for the hierarchical structure and data content of the Prolog logic text.

Chapter 10 contains an experiment as part of the demonstration case study. For conciseness, a simple referent is selected. It is first described in natural language, and is then specified in each specification medium in turn. Each specification is translated into the reference language Prolog, in preparation for measurement. The resulting measures are summarized in the chapter, and the detailed measures are shown in the Appendix.

References

1   Jevons, WS (1873): 'The Principles of Science" republished by Nagel, E (1953) Dover Publications.

2   Bunge, M (1967): "Scientific research". Springer Verlag.

3   Kyburg, H E (1984): "Measurement theory". Cambridge University Press.

4   Krantz, D H, Luce, R D, Suppes, P, Turski, A (1971): "Foundations of measurement". Addison Wesley.

5   Katzner, D W (1983): "Analysis without measurement". Cambridge University Press.

6   Roberts, F S (1975): "Measurement theory, with applications to decision-making, utility and the social sciences", Addison Wesley.

7   Kaposi, A A (1991): "Measurement theory". Ch. 12 in "The Software Engineer's Reference Book", ed. J McDermid, Butterworth Heinemann .

2 SYSTEMS AND MODELS

2.1 Intuitive notions

What is a system?

The dictionary definition of 'system' is "a whole, composed of parts in orderly arrangement according to some scheme or plan"([1]). This emphasizes two attributes: a system is a *unity*, and it is a *composite*. In accord with systems theory (see e.g. [2], [3], [4]), we regard 'system' as a generic notion which applies to any entity: concrete or abstract, natural or man-made, alive or inanimate. Any entity may be considered as a unity distinguished by the combination of its attributes, or as a composition of parts.

Why model?

The need for models arises because any real-life system, even a simple everyday object, is inherently complicated. It is impossible to comprehend fully the intricate interaction of any entity of the real world with its environment, or to describe all its attributes and each of its detail. Models are manageable simplifications, especially devised to aid understanding. They draw a boundary around the object of concern, defining it and separating it from its context. Models can describe what exists, specify what is required, and facilitate the design of future systems suited to fulfil the need.

Models and their uses have been the subject of intensive study by systems theorists. In one of his books which has since become a classic of cybernetic systems theory, W Ross Ashby ([5]) illustrates the need for models in the following passage:

> "*Every material object contains no less than an infinity of variables and therefore possible systems. The real pendulum, for instance, has not only length and position: it also has mass, temperature, electric conductivity, crystalline structure, chemical impurities, some radio activity, velocity,*

reflecting power, tensile strength, a surface film of moisture, bacterial contamination, an optical absorption, elasticity, shape, specific gravity, and so on and so on. Any suggestion that we should study "all" the facets is unrealistic, and actually the attempt is never made. What is necessary is that we should pick out and study the facts that are relevant to some main interest...".

Here Ashby looks at his pendulum as a *unity*. He might also have explained the inherent structural intricacy of the *composition* of the pendulum, apparently made up of only two constituents: a string and weighty object, each one decomposable into its material fibres or crystalline structures, atoms and subatomic parts. While one must be aware of the underlying structure and be ready to explore it, for practical purposes most of the detail is irrelevant. The function of models is selectively to represent the important features of the infinity of detail. The most helpful model is the simplest which still suffices.

Models may be put to many uses, among these:

- to describe some existing system clearly and concisely;
- to prescribe the main characteristics of some future system;
- to communicate system properties effectively;
- to reason closely about a system, and analyze it rigorously;
- to predict system properties reliably;
- to select an existing system to meet requirements, and choose judiciously among options;
- to design a system to given requirements;
- to set examples and norms.

Who needs models?

Models are needed by us all when we wish to describe any entity: a natural habitat and the organisms living within it, an industrial organization and its members, a product and the processes which give rise to it, objects, services and phenomena of everyday life, such as foods, tools, houses, train timetables, electricity bills, the weather and the passing of seasons. In the past, the study of models and modelling has been the domain of some selected classes of professionals, such as scientists, economists, engineers and mathematicians. Now that the 'computer revolution' facilitates the creation of systems of unprecedented power, and such systems are put into use to control a widening domain of human activity, the need to understand the principles of making, using and judging models has become a matter of general concern. Models have become a topic of interest among doctors and teachers, bankers and administrators, businessmen and managers, suppliers and end-users, children and adults, creaters and enforcers of standards and the public at large.

What is a model?

The everyday meaning of the word 'model' is defined in the dictionary ([1]) in two ways. A model may be *'an object of imitation'*, such as a person who poses for artists, a role model, or some 'exemplar of excellence'. A model may also be a *'representation'*. In this sense, a model may be a design for a new project; a template or prototype; a mould; a drawing; something that resembles something else.

In this book, we use the word 'model' to designate a type of representation. In practical situations, models are purposefully simplified representations of one entity by another. The original entity of interest is called the 'referent'. Its representation is the 'model'. The simplification is achieved by selectivity: attributes of the referent which are essential to the given purpose are preserved in the model; others are deliberately suppressed. It is often helpful to restrict models to finite systems, characterized by a finite set of properties and composed of a finite set of parts ([6]).

In our usage, 'model' is a relative notion: in itself, the statement "X is a model" is meaningless. The statement "X is a model of Y" is meaningful, but incomplete. The statement "X is a model of Y with respect to attributes a, b and c" is meaningful and complete, but not necessarily true, because the model may or may not be a valid representation of the referent.

What is a 'good' model?

Because of the infinite richness and variability of real life, any entity can give rise to an infinity of different models, each a valid representation of the referent. However, to be of practical value to a specific class of user, and to serve a given user's distinct purpose, each model may have to preserve a different collection of the attributes of the referent. This is why selectivity is essential to modelling. The dependency of the choice of model on the user and his/her needs is sometimes referred to as the 'relativistic property' of models.

Modelling involves error. Since a model displays only a selection of the infinity of attributes and parts of its real-life referent, it always deviates from its referent. Moreover, the model cannot reproduce the selected characteristics with complete precision, and even if such perfection could be achieved, we would have no means to detect it. Thus, the model must always be seen as an approximation, representing the selected characteristics approximately, with finite accuracy, but within defined bounds of error.

The 'good' model conveys to the user the characteristics chosen by the modeller. The medium of representation (the 'language' of the model) must meet three requirements: it must suit the modeller – the sender of the message; it must suit the referent in its context, capable of capturing its chosen characteristics so as to carry the message; it must suit the intended user – the recipient of the message.

When judging models, the notion of 'quality' (fitness for purpose) applies. A 'theory of models' allows us to identify some general properties common to all sound and valid models. In addition, a model must fulfil some specific requirements, imposed by the given referent and the given situation.

Models: benefits and dangers

Good models aid understanding, support communication, facilitate scientific advance, and serve technology. They help members of the public to identify products and services relevant to their needs, assist transaction of business, and provide protection of the public interest through the formulation of standards.

Poor-quality models fail their users. There are many ways in which a model may be deficient; for example:

- The model may be *ambiguous*, or even *meaningless*, if its referent is not clearly designated or its attributes are not properly defined.
- The model may be *invalid*, misrepresenting the referent.
- The model may be *incomplete*, describing some, but not all, of the required characteristics.
- The model may be *inaccurate*, failing to conform to set error limits.
- The model may be *redundant*, incorporating irrelevant features, thus distracting attention from the key characteristics.
- The model may be *unstable:* it may be valid at the time when it is created, but obsolete by the time it is used.
- The model may be temporally *incoherent*, in the sense that the various characteristics do not hold simultaneously.
- The model may be *incorrect* with respect to the syntax of the chosen medium of representation.
- The model may be *incomprehensible* to the user, describing the entity of interest in an inappropriate language.

In these and other ways, poor-quality models confuse rather than enlighten; they mislead their users, or overwhelm them with too much detail and bogus data. Appearing impressive and convincing, some poor models can inspire false confidence, but lead to wrong decisions and possible disasters. The user must be protected against such models by well-defined tests and procedures of quality assurance.

To summarize

Modelling is the purposefully simplified representation of one entity (the referent) by another (the model).

A valid model preserves selected attributes of its referent. Within finite error bounds, and with respect to the chosen attributes, it is a true representation of the referent.

To be judged of appropriate quality, a model must have an explicit referent, must preserve all chosen characteristics with acceptable accuracy, and must describe the referent in a language suited to the modeller and to the intended user of the model.

2.2 A systems approach to models

We adopt a *systems approach* as the foundation of a 'theory of models'. Such an approach is helpful in solving any kind of problem, but is a particularly useful source of organizing principles for coping with complexity.

- Any system may be viewed as an indivisible 'atomic' whole, and characterized accordingly, as a *unity*.

- Any system may be regarded as a *composition:* divisible into a structure of other systems which are its parts. Such a system can also be viewed as a *component:* part of a bigger, more elaborate system.

- The notion of *system* is used in the modelling process itself. Both the referent and its model may be regarded as systems, and the selected attributes define the relationship between the two.

2.2.1 'System': a formal definition

We define an arbitrary system S as a mathematical structure: the ordered pair

$$S = (\ E, R_E\ ) \dots\dots\text{Eq.2.1}$$

where $E = \{\ e_1, e_2, \dots e_n\ \}$ is a set of elements which may be finite or infinite,

$R_E = \{\ r_{E1}, r_{E2}, \dots r_{Em}\ \}$ is a set of relations on the elements of E, and equation 2.1 is referred to as the 'system definition' expression.

The alternative notation of equation 2.1a is also used sometimes, emphasizing that R_E imposes relations over the element set E:

$$S = R_E\ (\ E\ ) \dots\dots\text{Eq.2.1a}$$

The system definition expression is general, and makes explicit the essential distinction between a *system* and a *set*. Without the system-forming relation set, R_E, elements of the set E would remain disjoint. In a real-life system, elements of E can relate to each other in several ways. R_E carries the responsibility of providing the set of 'system constructor' relations which cement the elements of E into the coherent entity of the system as a unity.

2.2.2 A 'theory of models' in outline

Modelling is a process which forms a relationship between two systems, each definable in the form of equation 2.1. The referent system S may be a tangible entity: a single object or a finite or infinite collection. It may also be a natural phenomenon, or else some abstraction: a concept, a theory, or a computer program. The model system M is always a simplification, only preserving specifically chosen features of the referent. Models of interest to us here are finite systems. Frequently the referent is not an object of reality but is itself a model: the product of some earlier modelling process. In such cases the modelling process is a chain, each new link of which creates a model of a model, each new model being a further simplification of the previous.

In this section we sketch out a 'theory of models' in four stages. Sections 2.3 to 2.6 then elaborate on each of these in turn. We shall use the system definition equation 2.1 throughout. The theory explains the relationship between the referent, its model, its environment, its attributes, and its component parts.

STAGE 1 formalizes the relationship between the referent and its model. Assume that the referent system S belongs to the real world, called the 'WORLD OF THE REFERENT'. The system M is part of an artificial domain: the 'WORLD OF MODELS'. Let these two WORLDS form the elements of a system. The two elements are related in two ways: either by the *modelling process* which creates M from S by abstraction, or by the *instantiation* or *reification* process, moving from the general to the more specific, adding detail (Figure 2.1a). The notation $M(S)$ designates the relationship between the referent S and its model M.

STAGE 2 models the internal organization of each 'WORLD' as a system. In reality, the referent system will have a network of elaborate interactions with the numerous entities in its 'world'. In course of modelling, such interactions are simplified, defining only two systems: the REFERENT and its immediate ENVIRONMENT, which are in interaction as illutstrated in Figure 2.1b.

STAGE 3 focuses on the REFERENT, regarding it as a unity: an atomic entity. The referent is modelled as the system of related properties seen by an OBSERVER (Figure 2.1c).

STAGE 4 examines the REFERENT as a composition of PARTS: the way in which it is constructed from other systems – referents in their own right (Figure 2.1d).

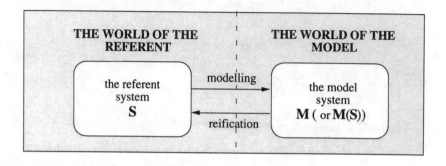

Figure 2.1a: Relationships between a systems of two 'WORLDS'

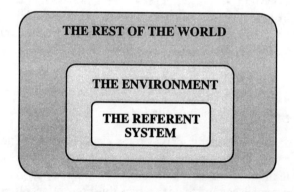

Figure 2.1b: The REFERENT and in relation to its ENVIRONMENT

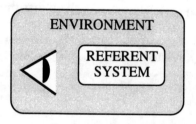

Figure 2.1c: The REFERENT observed in its 'WORLD'

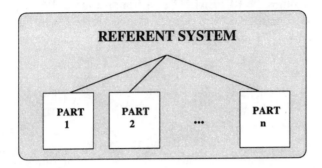

Figure 2.1d: Construction of the REFERENT from its component PARTS

2.3 The system formed by the referent and its model

Consider now some system **X**, made up of two elements: the referent system **S** and its model system **M**. The latter preserves some subset { **a, b,... z** } of the characteristics of the former. The referent is linked to the model by the relation 'is-modelled-by'. Using the format of system definition expression, introduced earlier, $S = (E, R_E)$ of equation 2.1, we now have:

$$X = (E, R_E),\quad \text{...Eq 2.2}$$

where $E = \{ S, M \}$,

and $R_E = r_{E1} = \text{is\_modelled\_by}$.

Let us examine the relationship between the elements of **E**. (The reader who is not familiar with the elementary notions of sets and relations is referred to introductory texts on discrete mathematics. A summary is given as mathematical background in several texts to which this book refers elsewhere, e.g.[7].)

(a)   Since the modelling process is a simplification, the process 'is_modelled_by' is ASYMMETRIC.

(b)   On the same basis, the relationship is IRREFLEXIVE.

Consider now an extension of our system **X**. The new system **X'** has three elements: **S, M** and **Q**, related together to form a 'modelling chain':

S is_modelled_by_M, and

M is_modelled_by_Q.

It follows from (a) and (b) that:

(c)   If M is a model of S and Q is a model of M then Q is also a model of S. Hence the relationship 'is_modelled_by' is TRANSITIVE.

The chain represented in Figures 2.2b can, of course, be extended.

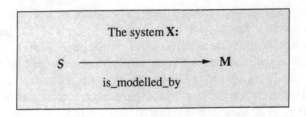

Figure 2.2a: The system **X**, formed by the referent **S** and its model **M**

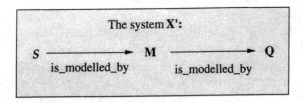

Figure 2.2b: The system **X'**, formed by the referent **S** and its models **M** and **Q**

Models, translations and designs

Suppose now that the entity Z^i in the modelling chain has some characteristics {**a, b, ..., z**} in common with the another such entity Z^j and is offered as its model. Three possibilities exist for representing a referent, as shown in Figure 2.3. Only in the first case is Z^i the model of Z^j.

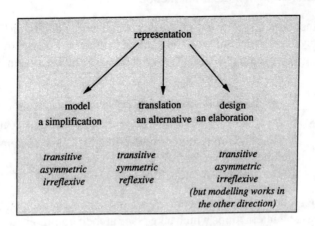

Figure 2.3: Model, translation and design

Models, designs and representations

It is often helpful to draw a strict distinction between 'model' and other forms of 'representation'. The model bears only the selected attributes of the referent, but a representation which is a design can have additional features of its own. Assume that **Y** is a design representation of a referent **S**, with respect to the attributes {**a**, **b**, ..., **z**}, and **M** is a model of **S** with respect to the same set of attributes. This means that **Y** is also a design representation of **M** with respect to those attributes (see [3]). We may write this formally as:

$$Y(M) = Y(S).$$

In this strict sense, Wren's timber scale model of St Paul's Cathedral is not a model but a design representation of the cathedral, because, although the real building and the scaled-down timber structure share an (abstract) model, and the two have sufficient common features for either to be recognizable from the other, both have distinct attributes of their own, outside the common model. Another example arises in the demonstration case study in Part 2 of this book, where the chosen attributes of some referent are expressed in a specification language, but the language has features of its own which are not present in the original referent.

2.4 The two-part system: referent and environment

Let us assume that a model is required to describe the relationship between the referent and the environment within which it exists (Figure 2.1b). The environment may be a single system, or a composite of many systems with which the referent is in mutual interaction.

In the first instance, one may define the entire WORLD OF THE REFERENT as a closed system, in accord with the general format of the system definition expression, given in equation 2.1. For this case, we write the system definition as:

$$W = (E, R_E), \dots\dots\dots\dots\dots\dots\dots\dots\dots\dots\dots\dots\dots\dots\dots\dots\dots\dots\dots Eq\ 2.3$$

where $E = \{ T, V \}$ refers to the world **W** as a two-part system, comprising the referent system denoted by **T**, and the environment denoted by **V**,

and R_E represents all interactions between **T** and **V**.

The two component parts of the world **W** are only aware of each other through observations and interactions which take place at their common boundary, and neither has interaction with any other entity. The structure of this two-part world is shown in Figure 2.4.

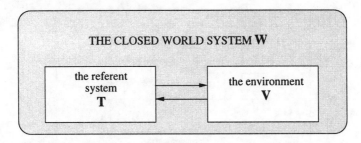

Figure 2.4: The referent and its environment in a 'two-part world'

Figure 2.4 has some notable characteristics:

(a) There is symmetry between **T** and **V**. While **V** is the context in which **T** exists, from **V**'s viewpoint one may also regard **T** as the 'context'. The two components of **W** are 'open' towards each other and are in mutual interaction. By contrast, **W** itself is a 'closed system', since it represents the entire WORLD. It can have no interaction with anything outside itself: its 'context' is empty.

(b) The two-part world of Figure 2.4 is so constructed that all stimuli and responses between the components should be exchanged through the boundary of **T**, which is also the boundary of **V**.

(c) We know that the closed system **W** can contain no element which is open from **T**'s viewpoint, other than **V**. However, **W** may incorporate any number of elements $U_1, U_2, ..., U_j$ which have no direct effect on **T** itself, nor any indirect effect on **T** through their interaction with **V**. This means that we may refine our initial model of the WORLD by drawing a tighter boundary around **V**, enhancing cohesion by forming a reduced 'active' environment **V'**). The remaining components $U_1, U_2, ..., U_j$ of **W** are also shown in Figure 2.5a. These are redundant from the viewpoint of the interactions between **T** and **V'**, and can be omitted from our simplified model of the closed world **W'**, now formed only of **T** and **V'** (Figure 2.5b).

(d) To assure the validity of this model, one must draw up a new version (eq. 2.4) of the system definition expression of equation 2.3, specifying the relations which must hold between the elements, such that the redundant elements may be omitted, while preserving the relation between the referent system **T** and its active environment **V'**.

The system definition is now:

$$W = (E, R_E') \quad \text{...Eq 2.4a}$$
where $E = \{ T, V', U_1, U_2, ..., U_j \}$.

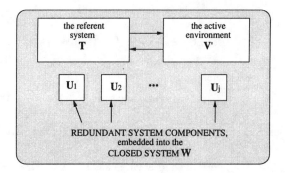

Figure 2.5a: Exposing the redundant parts of the 'world'

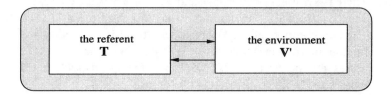

Figure 2.5b: Irredundant closed-world system, formed of the referent system and its active environment

The relationships between elements are expressed pairwise, therefore all the relations forming the set $\mathbf{R_{E'}}$ are binary:

$$\mathbf{R_{E'}} = \{\mathbf{R_{TV}}, \mathbf{R_{TU1}}, \mathbf{R_{TU2}}, ..., \mathbf{R_{TUj}}, \mathbf{R_{V'U1}}, \mathbf{R_{V'U2}}, ..., \mathbf{R_{V'Uj}},\}$$

Here $\mathbf{R_{TV'}}$ is a compound relation of the totality of interactions between $\mathbf{T}$ and $\mathbf{V'}$,

and $\mathbf{R_{TU1}}, \mathbf{R_{TU2}}, ..., \mathbf{R_{TUj}}$ explicitly state that there is NO interaction between the referent system $\mathbf{T}$ and any of the redundant components $\mathbf{U_1}, \mathbf{U_2}, ..., \mathbf{U_j}$.

Similarly, $\mathbf{R_{V'U1}}, \mathbf{R_{V'U2}}, ..., \mathbf{R_{V'Uj}}$ specify that there is NO interaction between the active environment $\mathbf{V'}$ and any of the redundant elements $\mathbf{U_1}, \mathbf{U_2}, ..., \mathbf{U_j}$.

It is of no consequence whether or not the redundant elements $\mathbf{U_1}, \mathbf{U_2}, ..., \mathbf{U_j}$ interact: these do not feature in the system definition of the irredundant closed world, which is now:

$$\mathbf{W'} = (\ \{\mathbf{T}, \mathbf{V'}\},\ \mathbf{R_{TV'}}\ ) \ ....\text{Eq 2.4b}$$

The system definition imposes no constraints on the interrelation between the redundant elements themselves. This means that these elements need not be closed systems: they may interact with each other, but this is of no concern from the viewpoint of $\mathbf{T}$ and $\mathbf{V'}$, and is not included in $\mathbf{R_{E'}}$.

The two-part model has a number of notable features. Its *simplicity* is a desirable characteristic which can be exploited throughout the processes of specification and design (as we demonstrate in Chapter 5). It is of particular value in analysing requirements for complex systems and drawing up their specifications. The simplicity does not mean over-simplification: as we noted before, the environment system **V'** may be a composite of many systems, each of which can interact with the referent system **T**. The *symmetry* between the referent and the environment is also a valuable attribute: by reversing the viewpoint, the methods used for modelling the referent may be employed for modelling its environment.

2.5 Black-box model: the referent as a unity

Let us focus now on the referent as an atomic system, indivisible and enclosed within a set boundary. We refer to the features of the referent as *attributes*, and to the characteristics of the model as *properties*.

The way in which the referent appears to an outside observer at its boundary (Figure 2.1c) is as a *black-box* or *behavioural* model. Such a model is called for by specifiers of requirements and end-users. The black-box model is also helpful to the problem solver in order to contain complexity by treating parts of the system as atomic.

The term *behaviour* refers to any characteristic of a black box, whether the attribute is static or dynamic, provided that it is observable without violating the boundary between the two parts of the closed system: the entity and its environment. (Note here that in some disciplines, such as computer science, the term 'property model' is preferred to 'behavioural model'; see e.g. [3]. Throughout this book we use the neutral terminology of 'black-box model').

Assume now that $M_B(S)$ (or simply M_B) is the black-box model of our referent S. Using the standard format of equation 2.1, the system definition now becomes:

$$M_B = (\textbf{Prop}, \textbf{R}_{\text{Prop}}), \dots\dots\dots\dots\dots\dots\dots\dots\dots\dots\dots\dots\dots\dots\text{Eq 2.5}$$

where M_B is the black-box (property) model of S,
and **Prop** = { $\textbf{prop}_1, \textbf{prop}_2, ..., \textbf{prop}_n$ } is the set of properties of the model of **S**: representations of **S**'s selected attributes,
and $\textbf{R}_{\text{Prop}}$ = { $\textbf{rp}_1, \textbf{rp}_2, ..., \textbf{rp}_m$ } is the relation set over **Prop**, forming M_B.

In Figure 2.6, we offer a hardware example where the referent is a simple commercially available device: a resistor whose pictorial model is drawn in the figure, actual size. Figure 2.6a shows a set of properties: the element set of a

Referent: the resistor R

The element set Prop = { prop₁, prop₂, ... propₙ }:

REF	NAME OF PROPERTY	DESCRIPTION/MEASURE
$prop_1$	type	carbon
$prop_2$	nominal value	1000 Ohm
$prop_3$	selection tolerance	$e = \pm 10\%$
$prop_4$	temperature coefficient	$k = 0.0015$ Ohm/°C
$prop_5$	maximum power dissipation	$P_{max} = 1/4$ Watts
$prop_6$	actual value (room temp)	1023 Ohm
$prop_7$	reliability	min. X hours to failure
$prop_8$	maximum acceptable operational value	$r_{op} = Y$ Ohm
$prop_9$	source of supply	Company XYZ
$prop_{10}$	price per 1000	£ Z
$prop_{11}$	length of body	$L = 3$ cm
$prop_{12}$	diameter of body	$W = 1$ cm
$prop_{13}$	volume of body	$V = 2.35$ cm^3
...		...
$prop_n$		...

Figure 2.6a: Example of a black-box model: property set of the resistor **R**

The relation set $R_{Prop} = \{\, rp_1,\, rp_2,\, \dots\, rp_m \}$:

REF	RELATION NAME	DESCRIPTION / MEASURE
rp_1	the 'individuality' relation	all n properties of the set Prop belong to the same referent, and are observed simultaneously, at a specified time
rp_2	the 'limit value' relation	$r_+ \le r_{actual} \le r_-$ and $r_{\pm} = r_{nominal} * (1 \pm e)$
rp_3	the 'operational value' relation	$r_{op} = r_{act} * (1 + k * T_c)$
rp_4	the 'volume' relation	$L * \pi * (W/2)^2$
...	...	...
rp_m	...	...

Figure 2.6b: Example of a black-box model: the relations

possible black-box model of the resistor. Figure 2.6b gives some of the relations which hold between the elements of the set. A black-box model of this kind may be of use to a variety of people, among them the manufacturer of this type of resistor, the distributor, the purchaser, and the potential users: the designer of electronic circuits, the production engineer concerned with implementation and the craftsman carrying out repair. Only a few of the resistor's black-box characteristics are listed in the figure: just sufficient to demonstrate the principles involved in the defining equation of black-box models. There is no claim that the model is, in any sense, complete; for example, the temperature coefficient is included but the voltage coefficient is not, although in different situations the two attributes may be equally important from the viewpoint of the circuit designer.

Figure 2.6b refers to rp_1 as the 'individuality relation'. This is a mandatory relation, to be included in all black-box models. It protects the meaningfulness of the black-box model against two kinds of errors:

- While a referent entity is characterized by its properties, a set of properties does not necessarily belong to the same entity. One of the functions of the individuality relation is to tie together a set of properties to form the black-box model of a specific referent.

- A surprisingly frequent modelling error is committed in some branches of technology by aggregating properties which are dispersed in time. Such a temporally incoherent model is liable to give a misleading impression, because there is no time at which one could guarantee that the set of property values correspond with those of the referent. An absurd example of this type of violation of the individuality relation would occur if a doctor would judge the state of health of the patient by considering his current blood pressure, his weight of last year, and his temperature last week when he had the 'flu, taking no account of the time displacement of the observed characteristics.

Accordingly, the individuality relation has two features:

- rp_1 synchronizes the property set, assigning to it a 'time signature';
- it holds over all of the property measures of the referent which are preserved in the black-box model. If the cardinality of the set **Prop** is **n**, then the individuality relation rp_1 is **(n+1)**-ary, where a place in the relation is given to each of the **n** members of **Prop**, in addition to the time signature.

The role of the individuality relation in assuring the stability of the black-box model of a given referent is discussed in further detail in Chapter 4, where we elaborate on the distinction between products and processes, and between individual products and product classes.

Key features of black-box models

Equation 2.5 and Figures 2.6a and 2.6b point to a number of important features of black-box models.

(a)   The property set **Prop**

This set represents each of the characteristics of a given referent by a measurable property. (Note that the criteria of measurability are set out in Chapter 3.) The set of property variables has n elements, of which some may be independent. In case of the example in Figure 2.6a, some of the independent properties of the resistor are: length, diameter, maximum power dissipation, and selection tolerance. Other properties are clearly interrelated, such as length and diameter with volume; price-per-1000 with selection tolerance; nominal value of resistance and temperature coefficient with actual value. The redundant information (e.g. the volume of the cylindrical body of the resistor, computable from its length and diameter) can be identified and eliminated from the property set. Alternatively, the redundancy can be put to good use, such as for checking the consistency of the data items of the black-box model.

(b)   The relation set $\mathbf{R}_{Prop}$

If the set **Prop** comprises orthogonal properties then the black-box model is irredundant, and the relation set $\mathbf{R}_{Prop}$ contains a single item: the mandatory individuality relation $\mathbf{rp}_1$. If **Prop** contains some inter-dependent properties then the relation set $\mathbf{R}_{Prop}$ must make these explicit in its further members. The interdependency relations in $\mathbf{R}_{Prop}$ can embody domain specific formal theories (such as the solid geometry of the cylinder, or the 'operational value' and 'limit value' relations in the example shown in Figure 2.6), empirical theories (raised pulse rate of the patient usually correlates with raised temperature), accumulated observations (scented melons are usually sweet and soft at one end), internally imposed constraints (operational value falling within the specified maximum selection tolerance of a resistor), and externally imposed constraints (such as standards).

2.6 Structural model: the referent as a composition

Structural models describe the internal organization of the referent, showing *what* parts it is composed of (Figure 2.1d), and *how* it is built up from its parts. Thus, the structural model is a system whose elements are the system's components, and whose relations are the structural interrelations between the components.

Although from the viewpoint of the end-user of a system structural models can be irrelevant, from the viewpoint of the designer, manufacturer, maintainer and other problem solvers concerned with the development, implementation and serviceability of the system, structural models are essential. They are the *means* of obtaining the desired black-box model of the referent. Structural models allow the deduction of the referent system's black-box model from the system structure and the black-box models of its component parts.

Consider now a referent system **S**, composed of **n** number of subsystems. Let $\mathbf{M_B(S)}$ (or simply $\mathbf{M_B}$) be the black-box model of the referent **S**, and let $\mathbf{M_S(S)}$ (or simply $\mathbf{M_S}$) denote the structural model of **S**, in which each subsystem is represented by its own black-box model. The general system definition of equation 2.1 is now formulated as follows:

$$\mathbf{M_S} = (\mathbf{Comp},\ \mathbf{R}_{Comp}) \dots\dots\dots\dots\dots\dots\dots\dots\text{Eq 2.6}$$

where **Comp** = { $comp_1$, $comp_2$, ..., $comp_n$} is the set of black-box models of the subsystems of which the referent is constructed,

and $\mathbf{R}_{Comp}$ = { $\mathbf{rc}_1$, $\mathbf{rc}_2$, ... $\mathbf{rc}_m$ } is the set of structural relations which hold over the elements of **Comp**.

ROW No	MODEL	COMPONENTS	STRUCTURAL RELATIONS
1	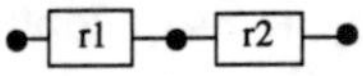	r1, r2	series
2	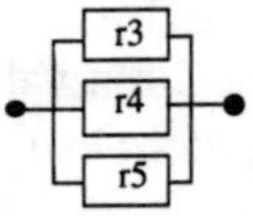	r3, r4, r5	parallel
3	a + b + c	a, b, c	add-to
4	d * e * f * g	d, e, f, g	multiply
5	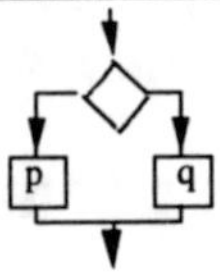	p, q	either-or (if-then-else)
6	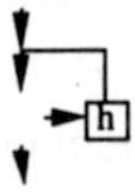	h	repeat-until
7		x, y, z	sequence
8	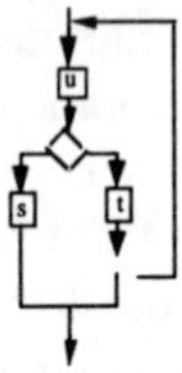	u, s, t	exit-from-the-middle
9	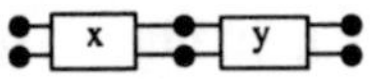	X, Y	concatenate
10	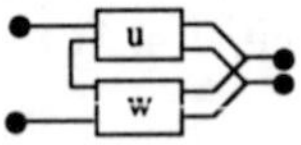	B, C, D	series-parallel

Predicates not labelled

Figure 2.7: Examples of components and relations of structural models

Figure 2.7 gives some simple and familiar hardware and software examples of components and structural relations of structural models. (We omit predicate names from flowcharts). Further examples are offered in [6].

2.6.1 Single-level structural models

Consider the single-level structural model of a system comprising **n** atomic subsystems. The component set **Comp** of the structural model will have **n** members, and the relation set R_{Comp} will comprise a single **n**-ary structural relation (equation 2.7). In such structures all components are peers. We sometimes refer to this type of structure as a 'homogene'.

$$M(S)_S = (\mathbf{Comp}, \mathbf{R_{Comp}}) = (\{\mathbf{comp}_1, \mathbf{comp}_{2, ...,} \mathbf{comp}_n\}, \{\mathbf{rc}_1\}) \ . Eq \ 2.7$$

where   **Comp** = $\{\mathbf{comp}_1, \mathbf{comp}_{2, ...,} \mathbf{comp}_n\}$ is the set of black-box models of the atomic components of which the referent is constructed,

and $\mathbf{R_{Comp}}$ = $\{\mathbf{rc}_1\}$ is the single-element structural relation set of the homogene.

According to this definition, row 3 of Figure 2.7, for example, is a homogene, defined as:

$$\mathbf{a + b + c} \ ,$$

or, using prefix notation for emphasis

$$\mathbf{+ (a, b, c)},$$

where the element set is **Comp** = $\{ \mathbf{a, b, c} \}$,

and   the structural relation set $\mathbf{R_{Comp}}$ comprises the single ternary relation $\mathbf{rc}_1$: the operation '+', defined over the element set $\{\mathbf{a, b, c}\}$.

Figure 2.8a is a *translation* of the homogene of equation 2.7: a directed graph representation which preserves the feature that **a**, **b** and **c** are the peer atomic component elements, and the root node carries the structural relation $\mathbf{rc}_1$, in this case '+'. The direction of the arrows in the figure shows the subordinacy of the component parts in relation to the root node, the whole of which represents the system.

The directed graph of Figure 2.8a may be generalized by removing the labels of the subordinate nodes. In this way, the graph becomes a *model*, capable of representing the sum of any three entities. As summation is commutative, the ordering of the leaf nodes is immaterial.

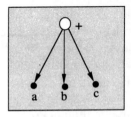

Figure 2.8a: Model of the homogene +(a, b, c)

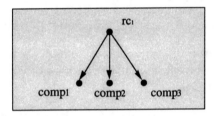

Figure 2.8b: The generalized tree of a ternary homogeneous structure

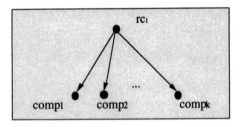

Figure 2.8c: The generalized tree of a k-ary homogene

The tree model does not display the specific character of the structural relation, but represents its existence by a symbol. The tree of Figure 2.8a can be generalized further by removing the specific label of the root node. With suitable relabelling, the resulting tree of Figure 2.8b would now be a valid representation of any of the ternary structures of Figure 2.7 (namely, those in rows 2, 3, 7), and may also describe other component referents, such as a small company with three employees, all reporting directly to the owner.

Trees are a popular and widely used representations of system structure. However, they must be applied with care, to avoid misunderstanding or unintentional loss of information.

- A frequent error is the casual omission of the structural relation, implicit in statements such as: "We solve the problem by decomposing it into parts." But how do we assemble solution from the part-solutions if the structural relation has been lost? "Our company consists of three departments." But how do the departments interact to perform the role of the company as a whole?

- Another source of difficulty is illustrated by the examples of Figure 2.7. The operations of summation (row 3) and parallel connection (row 2) being commutative, the ordering of the tree's leaf nodes has no significance; however, in other cases (such as row 7) the structural relation is not commutative, the position of the elements in the relation is significant in determining the outcome, and hence the generalized tree model of Figure 2.8b does not give sufficient information, even if the root label is preserved.

A further step of generalization leads to the tree representation of a **k**-ary (rather than specifically ternary) homogene, (Figure 2.8c). With suitable interpretation, this tree would now model several other rows of Figure 2.7.

The structural model of a single-level system

Assume that **S** is a system with black-box model $M_B(S) = (\textbf{Prop}, \textbf{R}_{Prop})$. Should **S** not be atomic, its black-box model is derivable from knowledge of its structure and the black-box behaviours of its components.

Let the structure of **S** be made up of two atomic components S_1 and S_2, with black-box models M_{B1} and M_{B2}, respectively. The structural model of **S** is $M_S(S)=(\textbf{Comp}, \textbf{R}_{Comp})$, where **Comp** is the set of black-box models of the composite **S**. Hence:

$$M_S(S) = (\{M_{1B}, M_{2B}\}, \textbf{R}_{Comp})$$
$$=(\{(\textbf{Prop}_1, \textbf{R}_{prop1}), (\textbf{Prop}_2, \textbf{R}_{prop2})\}, \textbf{R}_{Comp}).$$

This is depicted in Figure 2.9.

If $M_B(S)$ represents a desired black-box behaviour then its properties and relations must include *at least* those given in the original black-box model $M_B(S) = (\textbf{Prop}, \textbf{R}_{Prop})$. The structural model $M_S(S)$ is the designer's realization of that black-box behaviour, and it is likely that the black-box behaviour manifested by the finished system, $M_B(M_S(S))$, will show additional properties and relations not included in $M_B(S)$. We may therefore say that

$$M_B(S) \subseteq M_B(M_S(S)) , \text{ that } \textbf{Prop} \subseteq \textbf{Prop}_S , \text{ and that } \textbf{R}_{prop} \subseteq \textbf{R}_{propS}.$$

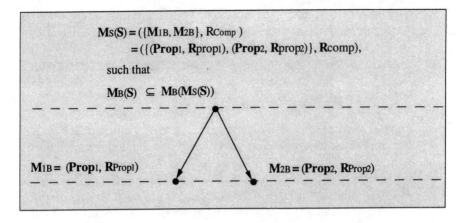

$$M_S(S) = (\{M_{1B}, M_{2B}\}, R_{Comp})$$
$$= (\{(Prop_1, Rprop_1), (Prop_2, Rprop_2)\}, Rcomp),$$

such that

$$M_B(S) \subseteq M_B(M_S(S))$$

$$M_{1B} = (Prop_1, RProp_1)$$

$$M_{2B} = (Prop_2, RProp_2)$$

Figure 2.9: The structural model of a single-level system

The system structure tree of single-level systems

Tree representations of structure, such as that of Figure 2.9, tend to be overloaded with meanings:

- the root node and the leaf nodes represent the black-box models of the whole system and its parts, respectively;
- the root node also represents the structural relation of the homogene.

The tree model is now expanded by adding a new root node which explicitly represents the black-box model of the system as a whole, confining the meaning of the old root node to the structural relation. Thus, for a three-element system, we extend the tree of the homogene of Figure 2.8a by the node **X** to represent the complete system **X** = **a** + **b** + **c** (Figure 2.10).

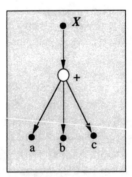

Figure 2.10: The tree model of the system $X = (\{a, b, c\}, \{+\})$

The tree is bipartite: there are two types of nodes. 'System nodes' or 'entity nodes' represent the black-box model of the complete system and its atomic subsystems. These are designated by solid dots. The 'structural relation node' of the homogene is designated by a circle. The two types of nodes alternate along the directed paths of the tree, and the graph of Figure 2.10 may be 'read' as follows:

> "**X** is the black-box model of a non-atomic system whose structure is composed by applying the relation '+' over the black-box models {**a**, **b**, **c**} of its atomic subsystems."

Of course, the bipartite tree of Figure 2.10 can be generalized for single-level structures with arbitrary number of subsystems. The tree classifies the systems of the homogene: the leaf elements are atomic, the root element is a composite, to be resolved into atomic elements by the structural relation. The relations of the bipartite tree are also classified: in addition to the **n**-ary structural relation applied over the atoms of the homogene, there is a unary nesting relation which identifies the system's black-box model with the structure.

2.6.2 Multilevel structural models

Most structures of practical interest are not homogenes: they are composed by nesting on several levels. Algebraic expressions would use a mixture of additions, multiplications, etc. Logical expressions would use both conjunctive (AND) and disjunctive (OR) composition. Two-port networks would use several types of interconnections. Since most equations will contain structures in a variety of combinations, even a relatively simple expression would amount to various homogenes nested on several levels. Similar examples are to be found in the fields of procedural programming, company structure, etc.

Such systems are *heterogeneous.* They are multilevel structures, composed by nesting together two or more homogenes. In some cases the inherent structure is obvious; in other instances superficial structural decomposition can lead to misleading results, and one must resort to sound domain-specific theories (such as classical network theory [8] or the theory of irreducible control flow-graphs [9]) to reveal it.

- Each homogene will have a single structural relation; however, the *component elements* of each homogene will be *classified:* some elements will be 'atomic', others 'structured' on further levels of detail, like the system itself, ultimately to be resolved into atomic components by lower-level nested homogenes.

- The *structural relations* will also be *classified:* some will be the **n**-ary relations of the homogenes themselves, others will carry out the function of nesting a homogeneous structure into one of the elements of another. Nesting is a unary relation, imposed on all elements which are non-atomic. These nesting relations, in effect, implement the classification of the elements of the structural model, identifying an element either as a structure or as an atom.

A two-level structure is given in Figure 2.11a, incorporating three atomic components (with black-box models M_{1B}, M_{21B} and M_{22B}), two sets of structural relations (R_{Comp} and R_{Comp2}), the intermediate non-atomic component (with black-box model M_{2B}) and the system itself (with black-box model M_B). The black-box model of the complete system is derived, level-by-level, in the manner shown in the previous case of Figure 2.9.

The system structure tree of multilevel systems

Expanding the model to represent multilevel systems in general, we designate the following steps of mapping for constructing multilevel trees.

(1) The black-box model of the system itself into the root of the tree, designated as a solid black node.

(2) The black-box models of the set of atomic elements into the leaves of the tree, all black.

(3) The black-box models of the set of intermediate elements into intermediate nodes, all black.

(4) The set of homogeneous structural relations into intermediate nodes of the tree, all white.

(5) Out-arcs from white nodes as pointers from the homogeneous structural relation to the elements over which it is defined.

(6) The single out-arc from a non-atomic black node as a pointer to the homogene structural relation by means of which it is composed.

The mapping extends the bipartite tree of the two-level structure shown in Figure 2.11a into the model of Figure 2.11b.

A fragment of the general bipartite tree is given in Figure 2.12, complete with its node labels. The structure starts from a black node: the root of the tree, representing the black-box model of the whole system. The structure continues on an arbitrary (finite) number of levels, finally terminating in the leaves of the tree: the black entity nodes representing the black-box model of the atomic components. The tree is bipartite: black and white nodes alternate along all routes, each white node representing structure.

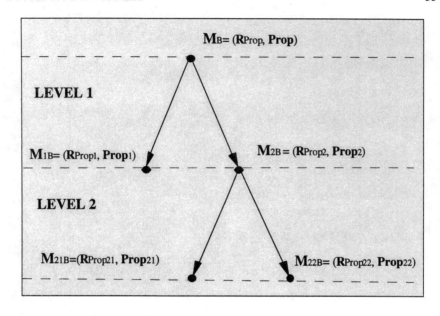

Figure 2.11a:
Deriving the black-box model of a two-level system from its structure

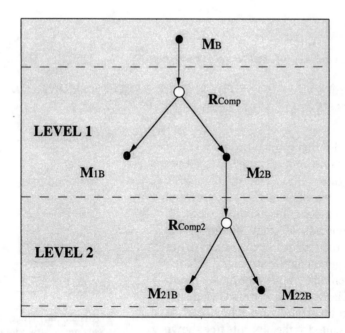

Figure 2.11b: The two-level bipartite tree of the structure of Figure 2.11a

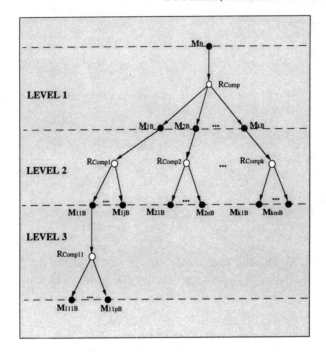

Figure 2.12: The general labelled bipartite tree

Simple examples of multilevel structures

FIRST EXAMPLE OF A TWO-LEVEL PROGRAM CONTROL STRUCTURE

Consider first a small program called 'C', whose control structure is shown in Figure 2.13a. The set of atomic assignment elements is $\{p, x, y, z\}$, and q is the sole intermediate element. Components of q are bound by the relation 'sequence'. The other homogene has the structural relation 'if-then-else', representing the predicate element. The system C is clearly a two-level structure which can be represented as the bipartite tree shown in Figure 2.13b.

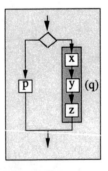

Figure 2.13a: The two-level control structure of program 'C'

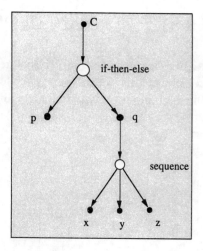

Figure 2.13b: Bipartite tree model of the program control structure 'C'

END OF FIRST PROGRAM CONTROL STRUCTURE EXAMPLE

EXAMPLE OF A TWO-LEVEL RESISTOR NETWORK

Consider now the two-level resistive network **Z** of Figure 2.13c. Exactly the same structural modelling process applies as previously. The three-component parallel homogene (r_3, r_4, r_5) is nested into r_2, and connected in series with r_1 to form **Z**. The bipartite tree model is the same as that of Figure 2.13b, suitably re-labelled.

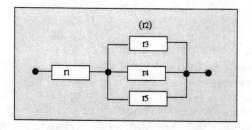

Figure 2.13c: A two-tier nested structure of resistors

END OF RESISTOR NETWORK EXAMPLE

SECOND EXAMPLE OF A TWO-LEVEL PROGRAM CONTROL STRUCTURE

Figure 2.13d shows a two-tier program control structure containing an 'exit from the middle' construct. This system will generate the same bipartite tree as Figure 2.13b, again differing only in its labelling.

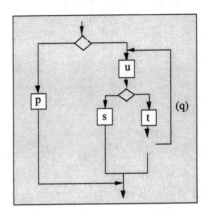

Figure 2.13d: A two-tier program control structure

END OF SECOND PROGRAM CONTROL STRUCTURE EXAMPLE

2.6.3 Abstractions on the structural model

It must be remembered that structural models are of secondary importance: modelling the system's black-box behaviour is always the ultimate aim. Nevertheless, structural models are powerful aids, and are essential in solving problems of complex systems. Structural models are often used without specific reference to black-box behaviour, for purposes such as:

- To represent the system and its parts symbolically.

  The general theories of uniport and two-port networks use structural models in this *abstract* way, and so do codes of practice of structured programming.

- To represent the sets of component parts of which a system is made, implying that they are structurally related, but not specifying the relations explicitly.

  We noted before that this type of *partial* structural model is frequently created accidentally, by those who fail to understand the distinction between a system and a set of unrelated parts. However, the partial model can also be constructed deliberately, to serve useful purposes, such as in parts lists for assemblies and subassemblies.

- To represent the set of relations which link component parts together, without making reference to the parts themselves.

  This second kind of *partial* structural model may be used in planning production processes, studying flowchart structures, or appraising the process chain to which a part is subject.

Each of these models is obtained by simplifying a bipartite structural model such as that shown in general form in Figure 2.12, subjecting it to one or more further modelling steps. We now show how to apply this modelling process to an instance of the trees of Figure 2.12.

The *first stage* of abstraction on the structural model consists of removing node labels. In doing so, we can identify a bag of leaf names, a bag of intermediate node names, and a bag of relation names as being removed from the model. The unlabelled tree created in this way is then a *generic* model, representing a wide range of systems. The equivalence (up to labelling) of the system structure trees of the three examples above shows this in Figures 2.13a, 2.13c and 2.13d.

The *second stage* of abstraction is suggested by the bipartite nature of the system structure tree, and consist of partitioning the graph, in each part retaining only one of the two types of nodes. The process is illustrated in the group of Figures 2.14:

- Figure 2.14a shows an unlabelled bipartite graph model, representing one of the multilevel trees of Figure 2.12.

- Figure 2.14b is the abstraction of a 'component tree', and

- Figure 2.14c is the abstraction of a 'process tree'.

The trees of Figure 2.14b and 2.14c are derived from the bipartite tree shown in Figure 2.14a by removing all nodes of one type with their in-arcs.

2.6.4 Quality of structural models

Examples such as those of Figures 2.13a to 2.13d draw attention to some important desirable properties of structures:

- Structures must be *realizable:* all of the components of the system must be either atomic elements, or intermediate elements resolved into structures of atomic elements.

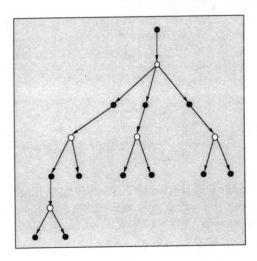

Figure 2.14a: An unlabelled bipartite tree

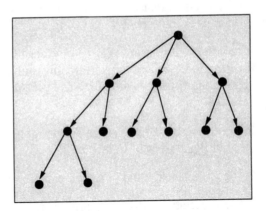

Figure 2.14b: Component tree

- All *atomic elements* of the structure must be *legal:* selected from a 'standard' set, where the standard may be 'globally' defined, as well as 'locally' constrained by the vendor and supplier companies.

- All *structural relations* should be *homogenes.* (If the user of the model fails to define the structure in terms of homogenes, it is the modeller's obligation to analyze the structure and resolve it into its inherent homogenes.)

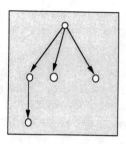

Figure 2.14c: Process tree

- All *homogenes* must be *legal:* the structural relations of the system should be choen from a 'standard' set, where the standard may be globally or locally defined.

In case of the simple examples of Figure 2.13, adherence to quality criteria can be readily checked by inspecting the system definition. For example, the designer of the resistive network would be advised to choose resistors r_1, r_3, r_4, r_5 from the international standard range of 'preferred values', assuring that the chosen values are consistent with company standards, and will be held in store. In case of the software example of Figure 2.13d, quality procedures will be applied to ascertain that the programmer has observed standards guidelines and house-rules of 'structured design' or 'structured programming' when using the exit-from-the-middle' structural relation. Additional quality criteria may be imposed by use of tools or procedures; for example, realizability will be checked by the compiler.

2.6.5 Constructing structures for complex systems

The management of complexity is one of the major problems of our times. The system definition expression of equation 2.1, and the black-box and structural models which arise from it, afford useful insight into the 'architecture' of complexity: the main causes of the difficulty of comprehending certain types of entities.

From our viewpoint, there are three factors which contribute towards system complexity:

- the 'size' of the system: the number of atomic elements to which the system resolves;

- the 'structure' of the system: the number and characteristics of the system-forming relations required for constructing the system from its atomic elements;

- the 'variety' of the elements: the number of different black-box models and modelling 'paradigms' required for characterizing the system and its elements.

In our simple examples system size was modest, and hence it was possible to define the system structure in terms of the complete set of components (atomic and intermediate), and the complete set of structural relations (on all levels of the hierarchy). This form of system definition facilitated the quality checks just described.

In case of any sizeable multilevel systems, it would be impractical (or even impossible!) to generate such an explicit and comprehensive system definition. Instead, elaborate multilevel structures of complex systems are defined recursively, or built up as nested substructures, with minimum variety of homogenes. The substructures are subsystems, and the bipartite tree model lends itself readily to their definition as subtrees. For example, one of the subsystems of the 3-level system structure in Figure 2.12 (M_{1B}) is a two-level structure in its own right, containing a one-level structure (M_{11B}) which finally resolves to atomic elements (M_{111B}, ...,M_{11pB}). The tree model of the subsystem is that of Figure 2.15.

The representation of subsystems as subtrees illustrates the principle of systems theory that any system can be a component when viewed from above, and any component can be a system when viewed from below. In the bipartite tree, all the internal entity nodes, such as M_{1B}, have this dual character: they may all be viewed as either component elements of the higher-level structure, or as structures containing lower-level components. The quality criteria that apply to systems also apply to subsystems. If the system is 'complex', comprising many varied elements and elaborate structural relations, it is essential, but not sufficient, to employ appropriate retrospective quality procedures and tools to *check* that the structure which has been created is of adequate quality. Design guidelines and tools are also required which assure quality constructively, to *prevent* the creation of poor quality structures, and actively *stimulate* the generation of structures of adequate quality.

Note finally the great practical value of the system principle just described. It allows flexibility in the definition of the atomic elements of the structure. The quality criterion of 'realizability' demands that the designer should take due care to assure that at any instant a comprehensive set of atomic components exists, into which the structure is resolved. However, design is always open-ended, involving invention, judgement and choice, and realizability cannot be guaranteed. If it turns out that no suitable ready-made component can be found for an element which was previously assumed 'atomic', then that element must be refined into a structure and made up of parts, so as not to disturb the design of the rest of the system.

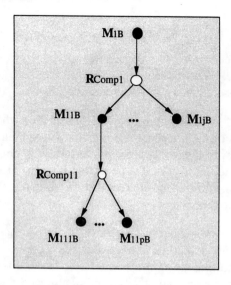

Figure 2.15: Bipartite tree model of a subsystem

2.6.6 Some key features of structural models

Alternative structural models

Structural models are not unique. Just as one can construct any number of equally valid black-box models of a given referent, there are many different ways in which a referent may be built up of parts, depending on the modeller's viewpoint and resources. The variety of structural models can be constrained by keeping the atomic elements invariant; however, since the same set of atomic elements may enter into a large number of different but valid and revealing structural relations, a virtually unlimited variety still remains.

Consider for example the structural modelling of some industrial product, made up as a multilevel assembly of 'ready-made' atomic components. Technical departments will have to compose several different structural models of the product to describe its function, reliability behaviour, dimensions, physical layout, resource requirement, etc. Administrative, financial and managerial departments will also develop distinct structural models to represent the product's pricing structure, assembly progress, integration test procedures, etc. These models utilize different black-box features of the atomic and intermediate components, and impose over them different structural relations, from which the manufacturing company can obtain a sufficiently rich, comprehensive description of the behaviour of its product.

Trees and other directed graphs

In the whole of Section 2.6, trees emerge as fundamental, generic representations of structure. The tree describes the 'architecture' of the system: the interlinking of its elements and structural relations.

- The root of the tree symbolizes the system as a whole.
- The leaves of the tree stand for the atomic components.
- Other nodes of the tree represent intermediate component elements of the structure. Each black node of the bipartite tree nests one of the homogenes, and the out-degree of each white node preserves the arity of the homogene.

Because of their scope and expressive power, trees have been extensively used in many branches of science and technology to represent organizations, and people have found that the *shape* of the tree offers valuable insight into the manageability, efficiency and other quality features of the organization. The characterization of the shape of trees is one of the ways in which we illustrate model-based measurement in this book.

It should be borne in mind that various directed graphs, other than trees, also have a valuable role to play in modelling. To identify some of their underlying properties, such graphs can be converted into a tree by a further modelling step which removes some of the detail of structural information. As an example, consider a ring configuration of a network of workstations which is represented both by Figure 2.16a and 2.16b. The latter is an abstraction of the former, preserving the notion of the ring as a system of 5 connected elements, but losing the information as to which workstation is directly connected with which.

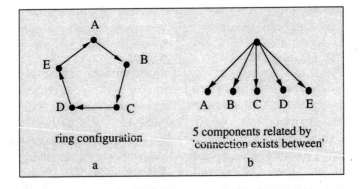

Figure 2.16: A network topology and its tree model

2.7 Summary

Problems and systems of the real world are complex. Models are needed to convey selected information, preserving primary characteristics, but also suppressing secondary attributes. Good models are purposeful, concise representations of entities of interest. They are the means of describing existing systems in interaction with their environment, they prescribe standards and specify requirements for future systems, and create new systems as structures of ready-made components so that they should be suited to fulfil specific needs. To be meaningful and useful, models must conform to general quality criteria. Among these are that a model must have a known referent, and the selected characteristics must be well defined and explicitly stated.

The generic notion of 'system' is a set of related elements. This is used to define:

- the relationship between the *'real world* of the referent system and the *'world of its models'*;
- the relationship between the *referent* system and its *environment:* the system with which it interacts;
- the *behaviour* of the referent as an indivisible whole, observed from the outside;
- the *structure* of the referent, which resolves the system into atomic parts, with the black-box behaviour of the referent arising from its structure and the black-box behaviour of its parts.

Black-box models are the means of describing a system as a unity. The elements of the black-box model of a given referent are a set of measurable selected properties. The meaningfulness of the black-box model is guarded by the individuality relation which assures that the properties belong to the same referent, and are synchronized, describing all the features of the referent at a given time instant. If the properties are orthogonal then the black-box model is irredundant, and its properties are linked only by the individuality relation. If the properties are interrelated, then the black-box model contains a set of relations which describe these interdependencies. Such black-box models are redundant, but the redundancy can be put to good use in assuring the consistency of the model.

It is not always possible or convenient to obtain the black-box model of a system directly. Structural models are the means of deriving black-box models from knowledge of the structure of the system and the black-box model of its atomic component parts. A single-level structure comprises a set of atomic elements, linked together by a single peer relation. Multilevel structures are also resolvable to atomic elements, but they also contain intermediate elements, composed, level-by-level, by nested peer relations. The bipartite tree emerges as a useful medium of structural modelling. Engineering practice, both hardware and software, has shown that simple graph structures tend to lead to 'sound' designs and successful products, whereas complicated constructs are conducive of design error. The methodological trends of 'structured design' in procedural programming, and the corresponding de-facto design standards of the IT industry, have been initiated by intuitive studies of abstract software structures, followed later by theoretical and methodological developments, and the construction of measures and tools for restructuring, maintenance and 'reverse engineering'. In Part 2 of this book we shall indicate that structural models and their measures can also throw light on the properties of declarative languages, and the specifications they describe.

2.8 References

1   Shorter Oxford Dictionary, 1987.

2   Blauberg I V, Sadovsky V N, Yudin E G (1977): "Systems theory". Progress Publishers, Moscow.

3   Emery F E (1969): "Systems thinking". Penguin.

4   Umpleby A S, Sadovsky, V N (1991): "A science of goal formulation". Hemisphere Publishing, New York.

5   Ross Ashby W (1956): "An introduction to cybernetics". Methuen.

6   Kaposi A A, Pyle I (1993): "Systems are not only software". Software Engineering Journal, Vol. 8, January, pp 31-39. IEE.

7   Sowa J F (1984): "Conceptual structures", Addison Wesley.

8   Hayt W H, Kemmerly J E (1987): "Engineering circuit analysis". McGraw Hill.

9   Fenton N E (1991): "Software metrics". Chapman and Hall.

"Nothing exists in itself. There is no quality in this world that is not what it is merely by contrast." Herman Melville, Moby-Dick

3 MEASURES

3.1 Introduction

This chapter sets out the concepts, principles and theoretical foundations of measurement, and defines the criteria of measurability. Modelling as the foundation of measurement is emphasized throughout.

Why measure?

The need for measurement arises in all kinds of situations: science and business, work and leisure, professional activity and daily life. Circumstances focus attention on some entity – the referent – which is of concern. Of the countless attributes possessed by any real-life referent, the given problem highlights one or more features of specific interest. The solution of the problem will call for a decision, such as choosing between items of a given class, or judging whether or not a given item meets set criteria or conforms to standards. Measurement assists the decision maker in problem formulation, observation, informed comparison, and reasoned discrimination.

Referents, attributes, properties and measures

The REFERENT of concern may be a tangible entity: a motor car, a person, a family, a piece of metal, an item of clothing. The problem may also relate to a phenomenon, or an abstract notion: the weather, the theory of motion, a computer program, the colour scheme in a room. The referent may be a single item (with some standard item in the background, to afford comparison), or a whole class of entities.

A given problem will call attention to one or more ATTRIBUTE of specific interest. The attribute must be related to some well-understood concept, and defined in a PROPERTY: a variable representing the attribute for all entities for which the concept has meaning. The attribute, and the associated property, may be *objectively observable*, such as the weight of a person or other material object, the cost of an artefact or a service, the number of statements or the number of lines of code of a Cobol program. Properties may also have to be defined for attributes which are *subjective*, such as charm which a person possesses, the value for money which an artefact may represent to someone, and the appropriateness of the programming style of a COBOL program.

A MEASURE is a particular value of a property variable. To characterize a referent, a measure must be assigned to each property which corresponds to its (objective or subjective) attributes.

Measures may be *quantitative* or *qualitative*. It is usual for objective attributes to be characterized by quantitative measures (e.g. 93 kg, £45.70, 8443 lines of code), but quantitative measures are also convenient for representing subjective attributes (3 marks out of 10 for charm, 7/20 for programming style, etc.). Here, the *symbol systems* in which the measures are expressed are the positive integers, the decimal fractions, and the vulgar fractions. Other symbol systems used are the Boolean or hexadecimal numbers, the complex numbers, arrays and matrices, etc. The choice of symbol system is guided by the nature of the problem, the competence of the measurer, the convenience of the user of the measures, and the required accuracy with which property values must be recorded.

Both objective and subjective attributes may be described *qualitatively*, for example, distinguishing people as fat/slim or charming/churlish, characterizing the cost of artefacts as cheap/medium-price/expensive, classifying COBOL programs as small/medium-size/large/extra-large, etc. In each of these cases the property variable has a small discrete domain (2 to 4 values), and the measure of a given referent would be one of these values. Symbols for the measures may be chosen from a range of numbers, letters, colours, shapes, or even pictures.

Quantitative measures consist of two parts: a numerical part, and a calibrating unit (kg, £), the *dimension* of the measure lending meaning to the numerical *magnitude*. It would be meaningless to say that the weight of a person is 93; the statements must be qualified by a unit of suitable dimension in which it is measured, (e.g. kg). Some quantitative measures are *scalars*, such as the ratio of two distances, the count of a number of items, or the radian measure of angle. We say that these quantitative measures have a *dimension of unity*. Whatever the dimension and the calibrating unit of the quantitative measure, in the definition of the measure it must be explicitly stated alongside the numerical magnitude, even when the measure is a scalar.

By contrast, qualitative measures are self-contained: they incorporate all the necessary information about the value of the property. We say that qualitative measures are *dimensionless*.

Measurement: process and measures

Figure 3.1 outlines the process of measurement. The referent resides in the REAL WORLD, where the 'empirical relation system' records observations about the selected attributes. The property variables which capture the attributes belong in the MODEL WORLD, as seen in Chapter 2. Measurement involves a third domain, because the measures are chosen from the WORLD OF SYMBOLS, where a 'formal relation system' represents property values and their interrelationships.

Measurement is then a two-step process:

- In STEP 1, a class of referents with a common attribute is *modelled* by a property variable, for the purpose of comparing items in the class.

- In STEP 2, a measure is *assigned* to the property variable of each item in the class, such that the comparison of measures should reflect comparisons about the referents.

Measures are the outcome of the two-step process leading from the referent via model. Figure 3.1a, as many others in this book, describes a simple or composite *process* which transform one *product* into another. Such 'product/process models' are presented in directed graphs which may assign a node to each product and an arc to each process, as in Figure 3.1a. Conversely, a directed graph may be composed which represents processes by nodes and products by arcs, and there will also be many instances of the use of that convention in this book. Wherever it is required for clarity, we show the key to the product/process model in the figure, but the discussion of product/process models is deferred until Chapter 4.

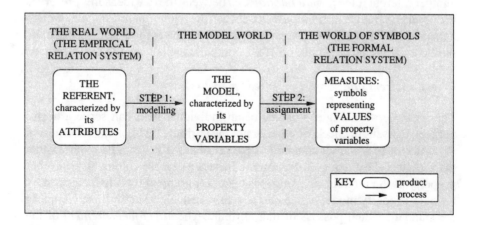

Figure 3.1a: Outline of the measurement process

We may now define 'measurement' and 'measure' as follows:

Definitions

Measurement is the process of making empirical observations about referent entities of the world, and representing their properties in a formal symbol system, so as to describe them.

A *measure* is the product of the measurement process. A measure is a symbol of a *symbol system*, designating the value of a property of the referent.

End of definitions

We may summarize as follows. Measures may be *quantitative* or *qualitative*. Quantitative measures have both magnitude and dimension. The *magnitude* of the measure represents the property by a specific symbol of the formal relation system of measurement. The *dimension* qualifies the magnitude of the quantitative measure by reference to the unit of measurement. Qualitative measures are dimensionless: they have magnitude only.

Meaning and validity of measures

A measure, by itself, is a symbol without context, and cannot have *meaning*. To have meaning, a measure must relate to a known attribute of a named referent by the process of interpretation (Figure 3.1b).

A meaningful measure may still not be *valid*. Measurement is a comparison, and hence the validity of a measure is relative. A measure of an attribute of the given referent is valid with relation to the measure of the same attribute of another referent if the *formal relationship* between the two measures preserves the observed *empirical relationship* between the two referents. In other words, a measure is valid relative to another measure if the comparison between the measures gives a true reflection of the comparison between the two referents.

Direct, indirect and 'object-oriented' measures

The measure of a single property of an individual item can serve a useful purpose in itself, to check whether or not the attribute fulfils some set criterion or conforms to some set standard. A packet of sugar is weighed to check that the weight is above some set minimum. In such a case the colour, taste or other property of the referent (the sugar) is taken for granted; it is characterized by a single-property model which can also represent all other entities, including the standard which possess of that attribute. The *direct measurement* of the property assigns a value to the referent, and also to the standard, such that the two may be compared.

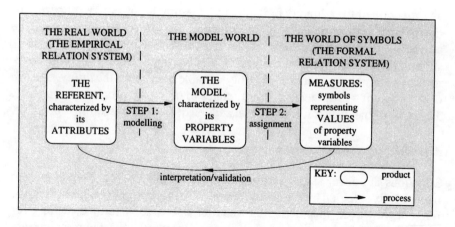

Figure 3.1b: Creating and interpreting measures

Often it is not feasible, or not convenient, to obtain the property measure of the required attribute by direct observation. Instead, one may have to collect property measures of several other attributes, none of which may be of particular interest in itself, but is related to the desired attribute in some known way. Using the appropriate relation of the black-box model of the referent, the required property can then be derived by *indirect measurement*. For example, it is convenient to measure the density of a lump of metal indirectly, by measuring, and suitably combining, measures of its mass and volume.

Many problems involve more than one attribute, and call for a more comprehensive, multi-property characterization of the referent. This is referred to as *object-oriented measurement*. Of course, the 'object' may be a material item, an abstract entity, a person, etc.

Utility measures

The ultimate aim of measurement is to assist *choice* and support *decision making*. Choice is always subjective, reflecting the preferences and value system of the decision maker, but this does not mean that the decision should be random or capricious. Rational, *informed* choice in science, technology or business relies on fact. Professional accountability requires that the factual basis of the choice should be defined, the value system be explicit, and the decision repeatable.

To reconcile the apparent conflict between the objectivity of observations and the subjectivity of value judgements in decision making, one must construct an explicit model of the subjective attribute on which the judgement is made, capturing it in a '*utility property*'. The model must describe the utility property as a *function* of the (directly or indirectly measurable) objective properties of the referent. The *arguments* of the function are the measures of objective attributes, recording the facts of the matter. The *form* of the function is subjectively determined by the problem solver/measurer, reflecting judgement. The function generates the utility variable. Its value – the *utility measure* of the referent – is subjective, but it relates to objective properties explicitly. The basis of the decision is traceable, and the decision is therefore repeatable.

A model-based measurement scheme

Situations are never static, and problems constantly evolve. The initial aim may be fully satisfied by the measurement of a single property of an individual referent; however, in practice the requirement is soon refined, calling for measurement of a wider range of attributes. Once collected, the measures are used for new purposes; yet further relevant attributes are identified and their measures obtained; utility measures are formulated to express and enhance decisions. Before long, the scope of the problem domain extends to a wider population of referents. This is the way in which insight into the problem domain develops.

To keep track of the growth of information, and to co-ordinate disjoint items of information into a coherent body of knowledge in the field, one needs a *system* which formalizes the relationships of attributes, properties, models and measures. We call such a system a *model-based measurement scheme*.

The rest of this chapter

Section 3.2 builds up a generic model-based measurement scheme, adaptable to a wide variety of measurement situations. We build up the measurement scheme in four stages:

(1) direct measurement of a single property,
(2) indirect measurement of a single property,
(3) object-oriented measurement, and
(4) construction of utility measures.

The ideas of model-based measurement are introduced informally, through simple everyday examples, where the measurement process is intuitive, and the solution to the problem is all but obvious. This allows the reader to grasp the principles of measurement without being distracted by the intricacies of the example itself.

Section 3.3 turns to *measurement theory* for explanation of the formal criteria which measures must satisfy: representation, uniqueness, meaningfulness and scaling.

Section 3.4 summarizes the theoretical and practical criteria of *measurability.*

Section 3.5 takes a broader view. It outlines the subject matter and structure of *'metrology':* the entire field of measurement, extending beyond the theory and practice of measurement to its legal implications and its role in technology development.

Section 3.6 provides some further examples, raising some topical issues relating to constructing model-based *measurement schemes for developing technologies* and new application domains.

3.2 Characterizing a referent by measurement

In this section we outline the measurement process through a series of examples.

3.2.1 Direct measurement of a single attribute

Our first example concerns a simple problem whose solution relies on the measurement of *width.* The empirical notion of width is taken as understood. It is captured by the well-defined property of linear distance, for which there are various agreed units of measurement.

GARAGING EXAMPLE

The problem

> The family Robinson live in an area of town where it is advisable to keep the car locked in a garage. The family garage is quite long, but very narrow. The Robinsons aim to choose the largest affordable car which could be parked in (or pushed into) the garage. Under consideration are the Metro, the Sierra and the Granada, listed here in increasing order of width (and hence in increasing order of family preference). The step-by-step process of measurement which leads to the purchasing decision is described below, and is shown in Figure 3.2.

The empirical relation system

> The referent comprises a set of four items:
> > { **U**(sable space in the garage); **M**(etro), **S**(ierra), **G**(ranada) } .
>
> The empirical relation of relevance to our problem is:
> > **M** is_smaller_than **S** is_smaller_than **G**.
>
> The problem is to find:
> > widest_of (**M, S, G**) is_not_greater_than **U**.

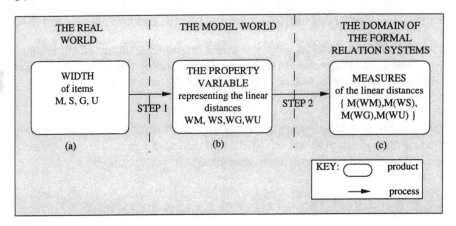

Figure 3.2: The measurement process of the GARAGING EXAMPLE

The model

The model suppresses all features of the referent, with the exception of the single property of **W**(idth), corresponding to linear distance. Each item in the referent is modelled by this property variable alone, and the set of elements of the model are:

$$\{ \, W_U, W_M, W_S, W_G \, \}, \text{ reading: "width of } U\text{", "width of } M\text{", etc.}$$

The empirical relations over the items in terms of their width can be formally expressed as:

$$W_M < W_S < W_G \dotfill \text{Eq 3.1}$$

The problem is to find:

$$\text{max\_of} \, (\, W_M, W_S, W_G \,) \leq W_U \dotfill \text{Eq 3.2}$$

The formal relation system

The measures of W_M, W_S, W_G and W_U are represented by $M(W_M)$, $M(W_S)$, $M(W_G)$, $M(W_U)$, respectively.

If all four measures are expressed in the same units (metres, inches, etc.) and differ only in magnitude, then relation (3.1) in the model maps into:

$$M(W_M) < M(W_S) < M(W_G) \dotfill \text{Eq 3.1'}$$

and the problem statement is now:

$$\text{max\_of} \, (\, M(W_M), M(W_S), M(W_G) \,) \leq M(W_U) \dotfill \text{Eq 3.2'}$$

The task is to find a formal relation system which is capable of representing the measures and their interrelationships.

Version 1a

Fenton ([1]) offers practical advice: try the natural numbers every time, and only choose something else if they fail. Let us adopt the set of natural numbers $\mathbb{N} = \{0, 1, 2, ...\}$ as the symbol system for describing magnitude of width, and let us choose the *inch* as the unit of measurement. We can now express, to the nearest inch, the four items of interest: the width of the three cars, and the usable internal dimension of the garage. For our particular referent (the Robinsons' garage and the types of cars), the outcome is shown in Figure 3.3. The message in the figure reads: "measure of the width of **M** is 64 inches", "measure of the width of **S** is 67 inches", etc., or, more concisely:

$$M(W_M) = \quad \textbf{64 inches,}$$
$$M(W_S) = \quad \textbf{67 inches,}$$
$$M(W_U) = \quad \textbf{68 inches,}$$
$$M(W_G) = \quad \textbf{69 inches.}$$

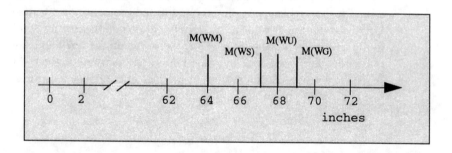

Figure 3.3: Measures of width of items of the referent, represented in the system of the natural numbers, and calibrated in inches

Substituting the measures into the problem statement of equation 3.2', we get:

max_of $(64, 67, 69) \leq 68$,

where all measures are dimensioned in inches.

This gives 67 inches: the width of the Sierra. Thus the solution of the problem, interpreted in the empirical domain, is that the Robinsons should buy the Sierra.

End of Version 1a

There is no doubt of the validity of the measures, but before acting on this solution, let us examine the quality of the result. Our selection of the number system was casual rather than deliberate, as was the choice of units and of accuracy of the measurement. Picking the inch as the dimension, and the natural numbers as the symbol system of magnitude, the measures are accurate to the nearest inch, but the problem seems also to be decided within an inch or so. We know from practice that measurement can never be perfectly accurate; could the inaccuracy of measurement lead to a wrong solution (the choice of a car which is slightly too wide), or could the solution be sub-optimal (a car which is not the widest possible)?

One way to improve the security of the solution is to re-examine the problem statement, given in the model by the inequality relation:

$$\text{max\_of } (M(W_M), M(W_S), M(W_G)) \le M(W_U) .... \text{Eq 3.2"}$$

There are at least three options for enhancing the solution:

- Tighten the constraint of the relation, replacing the weak order ($\le$) by the strict order ($<$). This protects against buying a car which is too wide for the usable space, but only a limited protection is afforded.
- Use the rounding error as a 'safety device'. When measuring width to the nearest inch, round *up* all the measures on the left-hand side of the inequality relation, and round *down* the measure on the right.
- Alter the formal relation system of the measure: choose a different symbol system, or a finer unit of measurement, or both.

Version 1b

The procedure and the final result are the same if we retain as the symbol system the set of natural numbers, but express width to the nearest centimetre, rather than to the nearest inch. The same queries arise about the quality of the result as before, but in this case the measures should give a 'good enough' solution: surely the Robinsons' problem cannot be so critical as to demand better than centimetre accuracy!

End of Version 1b

Version 1c

Let us measure in inches, but consider now that the formal relation system may have been ill-chosen, and may give too crude a measure. The problem is readily resolved by using as our formal relation system $\mathbb{R}$: the set of real numbers. These allow the width of items to be expressed to suitable fractional accuracy.

End of Version 1c

All the above versions had one thing in common: they used a formal relation system which mapped width into numbers representing width as linear distance (inches, centimetres, or other). This is a reasonable thing to do if the concept is well formed, and the unit of measurement is fully established, as is the case here. The four measures, read off a number line and calibrated in the same units of length, can do more than just *order* the four items in accord with their width: they also *quantify* width. From this information, one may deduce various new properties by calculation, such as the actual *width of the clearance* afforded by each car, the *average width* of the cars, the *percentage difference* between the widths, etc.

Although measurement always involves comparison (as observed by Melville), the calibrated measure gives a meaningful answer to problems, even when the referent comprises a single item, rather than a collection. The statement "the Sierra's width is 67 inches" is quite meaningful, even when no other item is measured. In this case comparison is not against other items of the referent, but against the unit of measurement itself.

Calibrated numerical measures are necessary for solving many problems, but for the purposes of the family Robinson these measures are too sophisticated. If the family were to be stranded without a tape measure or any other quantitative length-measuring device, their garaging problem could still be solved by use of a piece of string, with four knots carefully placed at points representing the widths of the three cars and the usable space. Let us forego the known units of length measure, and try for a really simple formal relation system which is still good enough in this particular case. What choices have we?

Version 2a

Try first the natural numbers, but this time on their own, treating them as qualitative measures, without attaching dimensions and units of length to them. The number line provides an infinity of choice of symbols: the integers from 0 upwards. We could use the same four numbers as before: 64, 67, 68 and 69, but now the statement "the Sierra's width is 67" would be completely meaningless: no individual measure would have any significance, only the relative *order* of the numbers. Our purpose would be served by any four distinct numbers. Let us then choose for our symbol set, arbitrarily, the small primes 3, 5, 7 and 11.

The knots on the string stand for the four widths. They follow each other in the order:

$$W_M, \text{ then } W_S, \text{ then } W_U, \text{ then } W_G.$$

The measures must preserve this order. When assigning symbols, we can exploit the ordering relation built into the system of the natural numbers: their sequence on the number line. The measures must be:

$$M(W_M), \text{ then } M(W_S), \text{ then } M(W_U), \text{ then } M(W_G).$$

This gives the measures:

$$
\begin{aligned}
M(W_M) &= 3, \\
M(W_S) &= 5, \\
M(W_U) &= 7, \\
M(W_G) &= 11.
\end{aligned}
$$

Since 5 is the largest number which comes before the limiting measure of the usable space (7), and since 5 is the symbol designating the width of the Sierra, the Robinsons' answer is the same as before.

End of Version 2a

Version 2b

Recall that in the qualitative formal relation system of Version 2a, only the order of the numbers was important: the numerical value had no meaning. Since one cannot take advantage of the calculating powers of numbers, is there any point in using numbers at all? In such a context, could the numbers be misleading, *implying* the possibility of drawing quantitative deductions, whereas none can be made? Could one find another order-preserving formal system which is simpler and more appropriate?

Any alphabet would serve as such a system. The Latin alphabet, used in the English language, has 26 upper or lower case letters – many more than the four distinct symbols required by the garaging problem. Let us choose – arbitrarily – the set of lower-case letters {**a, b, c, d**}. One of these must be assigned to the width of each of the items of the referent, such that their order in the alphabet should preserve the order:

$$W_M, \text{ then } W_S, \text{ then } W_U, \text{ then } W_G.$$

This gives the width measures:

$$
\begin{aligned}
M(W_M) &= a, \\
M(W_S) &= b, \\
M(W_U) &= c, \\
M(W_G) &= d.
\end{aligned}
$$

Here the boundary – the width of the usable space – is set by the symbol
c, and the nearest preceding symbol is **b**, standing as the measure of the
width of the Sierra. The Robinsons get the same answer yet again – a
reassuring confirmation of the validity of each version of our solution.

End of Version 2b

Observe that the measurement exercise yields two results:

- the *solution* to the problem as originally set out – namely, that the
  Sierra should be bought;
- a *measurement procedure* which could be used again were the
  problem were to be extended or modified, such as by adding some
  more car types to the selection.

END OF GARAGING EXAMPLE

Comments and conclusions on direct measurement of a single attribute

The example gives rise to the following general observations:

- Measurement always involves comparison. Items of the referent may be
  compared among themselves, or against a pre-set standard item, such as
  the unit of measurement.

- Measurement is a two-step process.

  STEP 1 maps the chosen, empirically observed attribute into a well-
  defined property, disregarding all other attributes.

  Note here the dependence of measurement on the quality of the
  definition and model of the property. In a mature discipline of science
  or technology, the attribute of interest is normally well understood, and
  the corresponding property is well defined. In new fields of application,
  and in developing technological domains such as software engineering,
  management science and information technology, the 'story' of the
  problem may bring to light new attributes which are yet to be fully
  understood. The concept must first be defined in words (see e.g. [2]), and
  represented in a property variable, as pre-requisites of proper measurement.

  STEP 2 maps the property of each item of the model into a symbol of
  the formal relation system. This involves a design element, because the
  formal relation system is not unique. Within limits, the measurer has
  freedom to choose between a variety of symbol systems and dimensions
  for representing the measures. The choice of the formal relation system
  must not affect the outcome: when interpreted in the empirical domain
  from which the problem arises, the result itself must be unique.

- Before acting on the result of measurement, the quality of the measures must be scrutinized to assure that it is valid, meaningful, not misleading, and neither too crude and error-prone, nor unduly complicated.

3.2.2 Indirect measurement of a single attribute

Indirect measurement relies on obtaining the measure of the required attribute through the measurement of properties corresponding to one or more other attributes. Kyburg [3] devotes great attention to the distinction between direct and indirect measurement, both from the practical and the theoretical viewpoint. He observes that "...the claim that a certain quantity cannot be measured directly is ambiguous": there is no known reason why a direct measurement procedure might not be developed in due course for all properties. However, at present relatively few properties are measured directly, because indirect measurement may be cheaper, more convenient, and often more accurate. Tried-and-tested tools and procedures are already available for measuring many properties, and can serve other purposes, provided that we can deduce other measures from them. Moreover, in some cases the scale of indirect measurement is more practical and more desirable.

Black-box models provide the basis of indirect measurement. We have seen in Chapter 2 that the black-box model $M_B(S)$ of referent S would be in the form:

$$M_B(S) = (\mathbf{Prop}, \mathbf{R}_{Prop}),$$

where $\mathbf{Prop} = \{ \mathbf{prop}_1, \mathbf{prop}_2, ..., \mathbf{prop}_n \}$ is the set of property variables corresponding to a set of attributes of S,

and $\mathbf{R}_{Prop} = \{ \mathbf{rp}_1, \mathbf{rp}_2, ..., \mathbf{rp}_m \}$
is the set of relations over $\mathbf{Prop}$.

$\mathbf{R}_{Prop}$ includes the individuality relation $\mathbf{rp}_1$, which assures that the property variables belong to the same referent and are measured at the same time. Other relations of $\mathbf{R}_{Prop}$ describe the interrelations among members of the property set. If now some required property $\mathbf{prop}_i$ is not directly measurable, but other related properties $\mathbf{prop}_j$, $\mathbf{prop}_k$, ..., $\mathbf{prop}_q$ are, then $\mathbf{prop}_i$ would be indirectly measurable, provided that it can be derived as some function $\mathbf{f}_i$ of the other, directly measurable properties:

Given $\mathbf{rp}_i (\mathbf{prop}_i, \mathbf{prop}_j, \mathbf{prop}_k, ..., \mathbf{prop}_q)$,
one may compute $\mathbf{prop}_i = \mathbf{f}_i (\mathbf{prop}_j, \mathbf{prop}_k, ..., \mathbf{prop}_q)$. ....Eq 3.3

EXAMPLE OF AREA MEASUREMENT – PART 1

We offer the simplest of all examples to illustrate the notion of indirect measurement. Consider the question of measuring the area of a page in a standard textbook. Intuitively, one would reach for a ruler (calibrated in centimetres, say), measure the length of the two adjacent sides of the rectangular page (22 cm and 14 cm, say), multiply the two measures together, and get the required area measure (308 cm^2). This is indirect measurement. The two length measures are of no interest in themselves, but are obtained as stepping stones towards achieving the sole required measure: that of the page area.

Could one measure area directly? What would one do in the absence of that knowledge of planar geometry which assures that the product of two orthogonal linear measures of a rectangle will be a valid measure of area? Given that we have a calibrated measure for unit area – such as a small square piece of paper, – one option is to cut out a sufficient number of clones of it, lay them on the page side-by-side and row-by-row, and start counting (taking care not to sneeze meanwhile). The other option is, of course, to develop a theory of planar geometry, and measure area indirectly.

END OF PART 1 OF AREA MEASUREMENT EXAMPLE

Direct and indirect measurement proceed similarly, and the schematic outline of Figure 3.1 applies to them both. However, indirect measurement has to accomplish a more complex task.

Assume that direct measures are needed for **n** auxiliary properties so as to generate the indirect measure for the $(n+1)^{th}$ property – the only one of interest. Then:

(1) The **n** auxiliary property measures must be obtained in accord with the procedure of the direct measurement described in Section 3.2.1 above, each yielding:

$$\mathbf{X}_i \rightarrow \mathbf{prop}_i \rightarrow \mathbf{M(prop}_i),$$

where $\mathbf{X}_i$ is one of **n** attributes characterized by direct measurement, **prop**$_i$ is the property which captures that attribute,
and $\mathbf{M(prop}_i)$ is the measure of the property.

(2) The empirical relation system must supply (or must have yielded before) a formal *theory*, built into the relation set of the black-box model of the referent. This models the attribute $\mathbf{X}_{n+1}$ by the property **prop**$_{n+1}$, and we compute it from the **n** directly measured properties in the manner just described.

The process is as given in Figure 3.4. All measures must have interpretation in the real world, and the interpretation arrow of the figure, linking the domain of the formal relation system to the real world domain, signifies this.

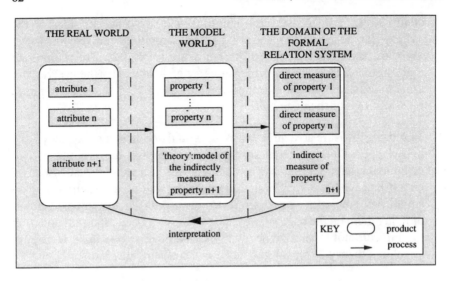

Figure 3.4: Introduction to indirect measurement

Let us review this process now for the area measurement example, following the previously described intuitive indirect measurement process. The example is too simple to illustrate all aspects of indirect measurement, but it shows the main features. Others will be featured in subsequent parts of the book.

EXAMPLE OF AREA MEASUREMENT – PART 2

The attribute set X has only three components: the length of the page X_L, the width of the page X_W, and the area X_A. The first two are modelled by the property of linear distance, and the third by area. The corresponding property set of the black-box model is {$prop_L$, $prop_W$, $prop_A$}. Only linear distance is deemed to be directly measurable. The black-box model will contain two relations: the individuality relation rp_1, and the ternary relation rp_2, with arguments ($prop_L$, $prop_W$, $prop_A$), embodying the theory of the planar geometry of the rectangle, which says:

> rp_2 is a relation over the properties $prop_L$, $prop_W$, $prop_A$, such that $prop_A$ is a product of $prop_L$ and $prop_W$.

The system of the natural numbers has symbols to represent the measure of $prop_L$, $prop_W$, $prop_A$. Assume that linear distance measures are calibrated in centimetres, the measurement yields the values $M(prop_L)$=22cm, and $M(prop_W)$=14cm. The area measure is calculated as the product:

$$M(prop_A) = M(prop_L) \times M(prop_W) = 22cm \times 14\ cm = 308cm^2.$$

END OF AREA MEASUREMENT EXAMPLE, PART 2

In this example, and in indirect measurement generally, the measurer has freedom to choose the formal relation system of each direct measure (within usual limits), but has no choice at all over the formal relation system of the indirectly measured property. In the present case, both parts of each direct measure (the magnitude and the dimension) enters into the formal relation which defines the indirectly measured property – in this case by multiplication. The product of the magnitudes of the direct measures gives the magnitude of the indirect measure ($22 \times 14 = 308$), and the product of the dimensions of direct measures gives the units of the indirect measure (cm x cm = cm^2). Without the latter, the former would be meaningless, and formally incorrect.

Since in this case both directly measured properties (length and width) *happened* to be measured in units of linear distance, it would have been contrary to common sense to choose different units for their measures (e.g. centimetres for length and inches for width). However, even for such an unconventional choice of dimensions, the area measure would have remained meaningful, and if the measurement had been conducted well, the result would have been the same. Let us check this briefly, using the real numbers to two decimal places as our formal relation system. Now:

$$\mathbf{M(prop_L)} = 22.00 \text{cm, and } \mathbf{M'(prop_W)} = 5.50 \text{ in.}$$

Hence the area measure becomes:

$$\mathbf{M'(prop_A)} = \mathbf{M(prop_L)} \times \mathbf{M'(prop_W)}$$
$$= 22.00 \text{cm} \times 5.50 \text{in} = 121.00 \text{cm-in.}$$

This can be readily converted into more convenient measures, such as cm^2, m^2, square-inches, etc. Choosing cm^2 for instance, we get:

$$\mathbf{M'(prop_A)} = \mathbf{M(prop_L)} \times \mathbf{M'(prop_W)}$$
$$= 22.00 \text{cm} \times 5.50 \text{in} \times \mathbf{\mu}$$
$$= 22.00 \text{cm} \times 5.50 \text{in} \times 2.54 \text{cm/in} = 307.34 \text{cm}^2,$$

where $\mu = 2.54$cm/in is the scale conversion factor from inches to centimetres.

(Note here that the use of mixed units has a hidden danger, because measuring distance to 2 decimal places in cm has a different accuracy from measuring it to 2 decimal places in inches.)

A glance at dimensional analysis

Observe that both the magnitude and the units of all quantities – the two direct measures and the conversion factor – entered into the calculation of the indirect measure, and the rules of arithmetic apply to both magnitudes and dimensions. The process of computing the dimension is sometimes referred to as 'dimensional analysis'.

Dimensional analysis is routinely used by scientists and engineers as an instrument of quality assurance. Its value can be seen even in case of our simple example. Elementary planar geometry demands that area measures should have the dimension of length square (whatever the units of length). If the dimension of the indirect measure deviates from this expectation then a logical error is detected in deriving the measure.

Consider now the case when the indirect measure is derived as the sum or difference of two direct measures. This was the case when measuring clearance as the difference between the width of the Robinsons' garage and of the parked car in the example of Section 3.2.1. Now, for the indirect measure for the clearance to be meaningful, all three measures must have the same dimension. Dimensional analysis flags an error if this invariance does not obtain.

In general, indirect measurement of a property involves the direct measurement of distinct attributes whose units of measurement therefore differ. This is the case, for example, when measuring velocity indirectly, through the direct measurement of distance and time, or when measuring electrical power dissipated in a resistor as a product of current and voltage, etc. The procedure of indirect measurement is the same in each case, and the same principles of dimensional analysis apply.

Summary and conclusions on indirect measurement

The description of the process of indirect measurement may be summarized with the aid of the measurement scheme of Figure 3.5.

- Measurement arises from some problem concerning the referent. The problem is usually described verbally, as a 'story', nominating the referent and the attribute of interest.

- Problem analysis reveals that there is no convenient measurement process for obtaining the required measure directly.

This raises two requirements for the *model* domain:

  * the definition and model of all relevant directly measurable properties;
  * the definition of the indirectly measured property, and a model of the property by use of a domain-specific *theory* which allows its derivation from directly measured properties.

Note that mature disciplines provide the necessary definitional and modelling foundations for measurement, but in new domains of technology and application these foundations are incomplete. They must be established before proper indirect measurement can be undertaken.

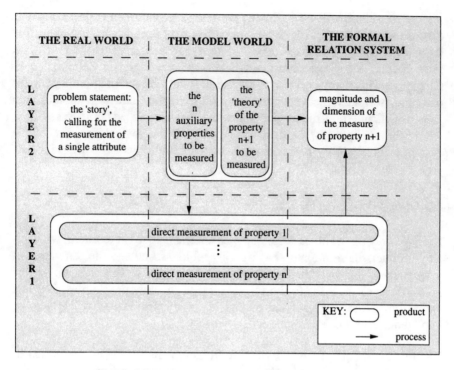

Figure 3.5: Indirect measurement of a single attribute

- In the *domain of formal relation systems,* the magnitude and dimensions of the indirect measure are derived from the direct measures which are combined in accord with the theory given in the model.

To keep the picture simple, Figure 3.5 omits arrows of interpretation, but of course all measures must correspond to attributes in the real world. Since the problem calls for the measurement of attribute **n+1** only, the 'owner' of the problem need not be burdened with detailed information about the other **n** auxiliary measures, and does not need to know the domain-specific theory for deriving from these the measure of the single relevant attribute **n+1**. The two-layered measurement scheme of Figure 3.5 separates the specification of the primary measurement problem (arising on LAYER 2) from the implementation of the solution (using the theories of LAYER 2, and relying on LAYER 1).

3.2.3 Object-oriented measurement

Most problems call for more detailed characterization of the referent than can be achieved by the measurement of just a single attribute. The black-box model is the basis of combining individual properties into a characterization of the referent as a whole. Some of the black-box properties will be directly measured; others will be indirectly measured in the manner just shown in Section 3.2.2. Yet other black-box properties may need to be computed from knowledge of the structure of the referent, and measures of the black-box properties of its atomic component parts, as described in Section 2.6 of Chapter 2. This many-faceted direct and indirect characterization is what we call object-oriented measurement. An instance is the characterization of both 'functional' and 'non-functional' attributes of an embedded computer system, where the attributes of interest extend beyond the logical function of the system to its reliability, performance, resource requirements, cost, ease of use, physical dimensions, etc.

Model-based measurement in research: identifying emergent properties

An important use of object-oriented measurement arises in research, when, as a by-product, the solution of the original problem stimulates the generation of a new scientific hypothesis. This process, which is fundamental to the scientific method, leads to the formulation of the new problem as a logical extension of the original, and results in the expansion and refinement of knowledge and the development of science.

To illustrate the point, consider that a problem calls for the measurement of some attribute A_x of the referent A, and A_x is indirectly measured, through the direct measurement of attributes A_y, A_z and A_w. The property variables are x, y, z, w, respectively. Assume that the model embodies a theory of the application domain which affords the calculation of x as:

$$x = y \times z + w \dots\dots\dots\dots\dots\dots\dots\dots\dots\dots\dots\dots\dots\dots\dots\text{Eq 3.4}$$

Analysis of the structure of equation 3.4 reveals the existence of some implicit variable $v = y \times z$. The variable is not some casual invention, nor does it come from statistical correlations over measured data. It arises naturally from the model itself. If the theory is trusted and well proven, the model will not lie; hence one may acknowledge that v is an 'emergent' property variable, pointing to some previously unobserved attribute A_v of the referent A. Research can then be directed at interpreting the meaning of the emergent property v in the real world of observable attributes, thus enriching understanding of the class of objects to which A belongs.

Note the value of dimensional analysis in this context. Assume that x is a variable denoting linear distance, measured in metres, say. Whatever the dimensions of y and z, dimensional consistency demands that their product, and w, should also be linear distances. When seeking to find a meaningful attribute corresponding to the new property variable v, the researcher needs only to consider those attributes of A which are measurable as linear distance.

Summary of object-oriented measurement

Object-oriented measurement involves (direct and/or indirect) measurement of each individual relevant property in the multi-faceted model, and the co-ordination of these by knowledge of the structure and use of the relations of the black-box model. Object-oriented measurement imposes an additional layer on Figure 3.5's layered process of indirect measurement. The measurement process is illustrated in Figure 3.6. As before, all measures must have interpretation in the real world, but interpretation arrows are omitted from the figure.

Let us examine once again the role of the supporting discipline in object-oriented measurement. A mature discipline of science or technology provides accumulated experience with the characterization of objects in the form of a

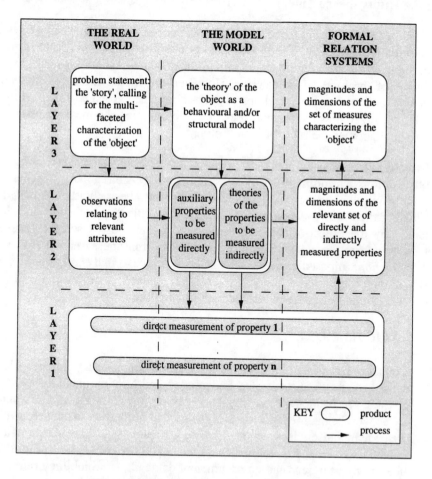

Figure 3.6: The process of object-oriented measurement

co-ordinated system of formal theories. Here the measurer's task is to make explicit the relevant parts of the wealth of theories which form the body of knowledge of the discipline domain. The theories combine the set of measures into a coherent and mutually compatible set of measures for the object.

When the discipline is still in its early stages of development, a comprehensive 'theory' of the object does not yet exist. The theory may lack the layer 1 foundation in a comprehensive set of directly measurable properties; it may not have adequate layer 2 support for deriving some of the required indirect measures; it may not have in place the layer 3 methodological apparatus for compiling the black-box / structural model of the object. Some form of three-layered formal theory must be constructed before embarking on object-oriented measurement. Otherwise object-oriented measurement is impossible.

3.2.4 Utility measurement

Consider again the multi-faceted model of the referent, obtained by object-oriented measurement. The model reflects the measurer's perception of the problem about the referent's most important attributes, and in this respect the *choice* of the model is subjective. However, the model itself provides an *objective* picture of empirical observations about the chosen attributes of the referent. It satisfies the formal rules of modelling, and the measure of each property satisfies the rules of measurement.

In many cases, measurement must go much further than objective recording of observations. The ultimate use of measurement is decision support. To achieve this, the problem solver's subjective judgement must be explicitly articulated. This is called *utility measurement*. The 'factual' objective measures are obtained by the object-oriented measurement process just described. These measures form the *elements* of a decision-supporting measurement system. The *functions* imposed on these factual elements give formal expression to the subjective criteria which model the decision maker's priorities and preferences. A simple example is included for illustration.

TRAFFIC EXAMPLE

Consider the problem of devising a scale of charges for a toll road which runs through the district of a fair-minded and 'environmentally aware' council. The council seeks to discourage heavy and noisy traffic, and wants to recover the maintenance costs of the road in proportion to the perceived wear, tear and environmental damage inflicted by each vehicle. After debate, a consensus is reached, and the council formulates the composite notion of 'wear, tear and environmental damage', to which they refer as 'WTE'.

It is realized that WTE relates to the weight of vehicles, but there is no facility to weigh the vehicles in transit. Instead, the toll charge is computed on the basis of measuring two properties: the vehicle's floor area, and noise. Floor area is indirectly measured, as a product of the direct measures of length and width (readily measurable by use of optical sensors). Noise is also measured directly, and recorded in decibels.

The utility variable which represents WTE will reflect the objective attributes of noise and floor area, as well as the aggregate views and preferences of the council. Let us assume that the council is uncertain as to how much to charge, but, as a matter of principle, it decides to make up the WTE charge as a sum of two parts:

- **P** pence for each decibel of noise (**N**), and
- **Q** pence for each m² of floor area (**A**),

where the exact magnitude of the two keys **P** and **Q** is yet to be settled by the council.

Then the utility model of WTE takes the form of the weighted sum:

$$k^{pence} = P^{pence/decibel} \times N^{decibels} + Q^{pence/m2} \times A^{m2},$$

and the utility measure of WTE is composed as a four-layer structure, shown in Figure 3.7 in the particular case. Figure 3.8 provides the general picture.

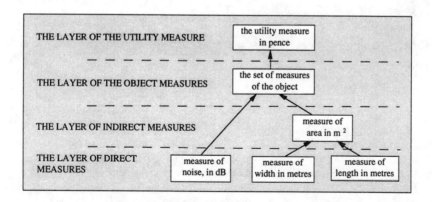

Figure 3.7: The structure of the WTE utility measure of the TRAFFIC EXAMPLE

END OF TRAFFIC EXAMPLE

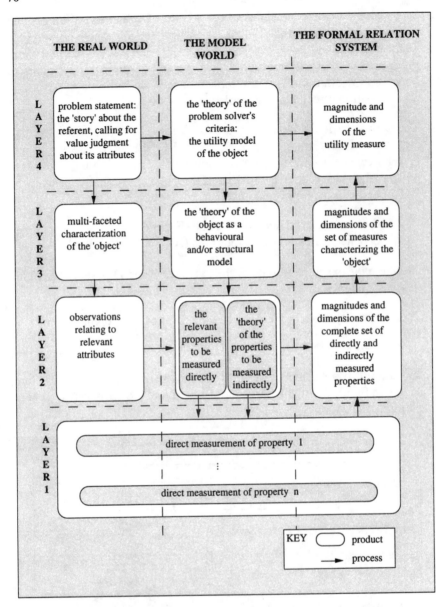

Figure 3.8: The process of utility measurement

Summary and conclusions on obtaining utility measures

Utility measurement builds on object-oriented measurement, imposing an additional layer, as seen in Figure 3.7, and as outlined in Figure 3.8. Here again, all measures must have interpretation in the real world, but interpretation arrows are omitted from the figure.

The 'story' of the decision problem leads to the utility model as the formal representation of the problem solver's criteria. This gives rise to the need for object-oriented measurement. The property measures of the object (directly and/or indirectly obtained) are pulled together by the subjectively composed utility model to yield the (magnitude and units of the) utility measure.

Figure 3.8 is a simplified representation. In practice the design of a utility measure is an iterative process, where the problem solver refines his/her judgement through experiments with various versions of the utility function. (In case of our traffic example, if the charging scheme fails to produce the required result, the council might try to assign differing weights to the factors **P** and **Q**, or may decide to use some other basis of imposing the toll charge).

In some instances a utility measure proves to be of use in a broader context than just a single problem, and may even acquire the status of a widely accepted measure of goodness: a 'Figure of merit'. For example, gain-bandwidth product and signal-to-noise ratio are universally adopted as two such measures of goodness of communication systems.

Note the important role of dimensional analysis in the design of utility measures. The 'story' of the problem determines the dimension of the utility measure (some practical unit of currency in case of the traffic example). In general, object-oriented measurement will contain measures of varied dimensions. The 'theory' of the utility measure must impose on these an appropriate relation such that the over-all dimensions should be meaningful. This may involve inventing constants which carry the appropriate dimensions. Dimensional consistency assures the quality of the utility measure, and must guide any evolutionary development.

3.2.5 A generic measurement scheme of model-based measurement

Figure 3.8 amounts to a generic measurement scheme: a system of model-based measures for objective and subjective characterization of the referent.

- Where the problem calls for decision making, the complete four-layer measurement scheme of Figure 3.8 is required. Any other measurement problem is a simplification.

- In the scheme, the measures are ultimately resolved into direct measures of the object. All measures are explicitly related to each other by models which gives the measures their context. A measure on a higher layer is derived from those on lower layers by a 'theory' expressed in the model. Once the model is devised and the measurement scheme is established, the model is implicit in all subsequent uses of the measures, and need not be specifically evoked in every case. (For example, the user of a thermometer need not know the theory of thermal expansion of liquids.)

- The direct measures of the scheme serve as a 'knowledge base' from which higher-layer measures are deduced. The characterization is not confined to the originally perceived attributes of the referent, but can be extended to new attributes whose importance may arise subsequently. The measurement scheme can cope with this by adding models at higher layers which derive the new indirect measures from the extant complement of direct measures.

- Evolutionary development of projects may call for extending the scope of measurement beyond the original set of direct measures. (For example, the original problem may have called only for characterization of the logical function of the referent – a computer-based system, – whereas later requirements may demand the measurement of non-functional attributes). This needs definition of further directly measurable properties at layer 1, necessitating the widening the model base at higher layers of the measurement scheme, while maintaining the measurement scheme's consistency throughout.

3.2.6 Validation of measures and measurement schemes

In principle, any entity can be modelled, and any model contains its own implicit measures. In practice, it may not always be possible or feasible to obtain measures which give an adequate impression of the referent entity under investigation. The reasons for such failure may be attributable to the lack of judgement of the modeller (the attributes selected for measurement may be useless from the viewpoint of the problem in hand, whereas the important attributes may have been omitted from the model). The fault may also be caused by the measure itself being invalid (mistakes may have been committed in either or both of the steps which lead from the referent to its measures as illustrated in Figure 3.1a). Failure of this second kind may be *prevented* by obeying formal criteria of measurability for each of the measured attributes. Criteria of measurability, discussed in Section 3.5, incorporate quality characteristics of the models and measures, and practical considerations relating to such factors as the resource requirements of the measurement process itself.

However trustworthy the quality assurance which guards the processes of modelling and measurement, all measures must be subject to validation against their original referent (Figure 3.1b). Well-trained young technologists are taught not to accept any measure without some form of independent check, even when the measured attribute is well understood, measuring instruments are well designed and properly calibrated, the measurement procedure is well established, and the units of measurement are well defined.

The burden of *validation of measures* in the field is much greater when the measurement is directed at some newly defined attribute where there may be insufficient reliable independent evidence against which to validate the measures. The difficulties are greatest when the measures relate to entities of some developing discipline domain, or emerge from some new field of application. It may take extensive investment, and many years of experimentation, data collection and interpretation, market research, scientific and technological debate, as well as argument and compromise in international committees, to validate a collection of measures, and to reach a consensus of their collective adequacy in characterizing some novel class of referent entities. In such cases the problems of the validation process are compounded by the uncertainties of the preceding processes of creating the models and measures.

Validation can be speeded up, validation costs reduced, and chance of failure minimized, by preliminary *validation* of the feasibility of the construction of a *model-based measurement scheme*. As previous sections of this chapter showed, the measurement scheme provides its own guards for the measure of each individual attribute. The models assure consistency between measures, defining their inter-relationships. Should validation of one of the chosen measures fail, others in the scheme might still prove valid. Work on obtaining the failed measure might not have been a wasted effort: the process may itself have contributed new insights into the characterization of some features of the referent, and may still be used as a constituent of some other, more useful measure. Finally, and most importantly, the continued validity of measures is assured by the dynamism afforded by the measurement scheme. As the discipline domain matures, extant measures may be enhanced, new measures may be developed, new attributes might emerge, and the measurement scheme may be enriched in the light of experience.

3.3 Characterizing measures

Measurement theory sets out criteria which measures must satisfy. The theory defines three such characteristics: *representativeness, uniqueness* and *meaningfulness*. These formalize the features of sound measurement which have been discussed informally in earlier parts of this chapter. In this context, measurement theory gives guidance to the design of *scales* of measurement, and we extend our discussion to these.

3.3.1 Representation

Consider again the measurement process outlined in Figure 3.1b. The process must yield measures which give a valid interpretation. This means that the measures effectively *model* the referent with respect to the chosen attribute, preserving the observed empirical relationships.

Measurement theory defines the formal *representation condition of measurement* as a modelling relation between a set of referents and their corresponding measures, which we have met in Chapter 2. Accordingly:

> A set of measures is a valid *representation* of a referent with respect to a given attribute if the mapping from the empirical domain of attributes to the formal domain of measures is a homomorphism: the relation is irreflexive, asymmetric and transitive.

Recall our 'garaging example', where the referent was a set of items, comprising the garage and a collection of cars, all characterized by the single observable attribute of width, and each item of the referent was assigned a single linear measure. Observe that the set of measures modelled the collection of referents, preserving their observed relationships with respect to width, and obeying the criteria of homomorphism.

3.3.2 Uniqueness and meaningfulness

Measurement codifies observations, so that the resulting measures should convey an unambiguous meaning. Measurement theory sets out a formal criterion of meaningfulness. It responds to the question: how can we choose a symbol which sensibly represents an attribute of a given referent? We noted earlier that it is meaningless to say that the weight of a man is 93, but it is meaningful to say that his weight is 93 kg, or 205 pounds. It is meaningful to say that there are 93 people in a meeting room. It may also be meaningful to say that the colour of a tin of paint is 93, but the meaning is not intuitively obvious, and meaningful interpretation of the statement would require reference to some previously defined colour code. Meaningfulness must not be confused with truthfulness or accuracy: it is meaningful, but not necessarily true, that there are 93 people in the room, and it is also meaningful to say that the weight of a person is 93 kg, even if his actual weight is 63 kg.

We have noted that the measurer frequently has a choice of symbol systems and units of measurement. The 'uniqueness condition' queries the range of options for the choice of symbols. Once a symbol system is chosen, can the scale of measurement be transformed into another scale, while preserving the meaningfulness of the measure, without invalidating the measurement? Measurement theory stipulates that, to guard the meaningfulness of a measurement statement, the scale must either be unique, or else we should select a scale such that a true statement should remain true. Formally, the interdependency of meaningfulness and scaling of the direct measurement of a single property may be defined as follows:

> A measurement statement expressed in a symbol of a formal relation system is *meaningful* if its truth or falsehood is invariant under all admissible scaling transformations between the original scale M and the transformed scale M'.

Recall again the 'garaging example'. The various 'versions' gave rise to the assignment of different sets of symbols to the set of referents, but the relationships between the set of symbols remained invariant, and the result – the choice of car – was unaffected by the choice of scale and scaling transformation.

Note that, in case of indirect measurement involving two or more directly measured auxiliary attributes, meaningfulness of the indirect measure relies on two kinds of criteria:

- each individual direct measure must be meaningful, and
- the modelling relation for the indirectly measured property must be preserved under all scaling transformations.

3.3.3 Scales and scaling transformations

We discriminate here five main scale types.

(1) The simplest form of scale is called the *nominal* scale of measurement. This kind of qualitative measure is needed when the empirical relation system calls for the *classification* of items of the referent by the chosen attribute. In case of nominal measurement, the only demand of the uniqueness condition is that the formal relation system should distinguish like from unlike, andshould contain a sufficiently large set of distinct symbols: if the referent has n items which are to be classified in m ways (where n m) then the symbol system must have at least m distinct symbols.

(2) The *ordinal* scale is used if the task is not just to sort items into groups, but also to order members of a group according to the extent to which they possess the chosen attribute. The measure is still qualitative, but the uniqueness condition demands that the formal relation system should be able to impose an appropriate *ordering relation* over the symbols. The alphabet is a popular and sensible symbol system for this type of measurement.

(3) The *interval* scale of measurement is quantitative. It allows the magnitude of the attribute to be expressed numerically, as a distance from some chosen point of reference. The formal relation system must be numerical, and must include the operation of difference. The integers and the real numbers are the usual choice of symbol systems, depending on the unit of measurement in which the dimension is expressed, and the accuracy of discrimination required.

(4) The quantitative *ratio* scale expresses the magnitude of the measure as a multiple of a chosen unit of measurement. The formal relation system must include sum/difference and multiplication/division. The real numbers provide the usual symbol system for this type of measurement.

(5) The quantitative *absolute* scale is reserved for counting, and uses the rational numbers as their unique symbol system of measurement. The unit of measurement is not negotiable because the attribute is resolvable to 'absolute' atoms, and the magnitude of the measure expresses the number of these.

The demands on these types of scales are not conflicting: they are cumulative, as one progresses down the list, the types of scales forming elements of a *system* of scale types, with an ordering relation imposed over them. Figure 3.9 summarizes the properties and uses of the five main types of scales, and the admissible scaling transformations within a given scale type.

The ordering relation over the scale types means that if a scale is suited for a higher-order purpose, it can also serve for all lower-order purposes. The converse is not necessarily true. Versions 1 and 1b of our 'garaging example' gave ratio measures of width, expressed in the symbol system of the natural numbers. From these, one may have drawn various quantitative conclusions, such as the width of clearance for the chosen car, the average width of the four cars, the extension of the garage which would be needed to accommodate the largest car, etc. Version 2 called for ordinal measures only. Although version 2a used the natural numbers, version 2b showed that the symbol system of the alphabet served the purpose equally well. However, the alphabet would have been inadequate for the purposes of version 1a, and could not have led to the kinds of quantitative conclusions afforded by the ratio scale measures.

Only in the case of absolute measurement is the scale unique. In all other cases it is possible to rescale the measures, provided that the new symbol system has at least as much expressive and manipulative power as the old. For example, one may first express the results of classification in a system of pictorial symbols, say, and the scale may then be mapped into a system of colours, the alphabet, or the set of natural or rational numbers. Some of these scales are also suited for higher-order measures, and their power is far from being fully exploited by the requirements of the classification problem. Of course, rescaling between formal systems of equal power is also possible, and rescaling is very often used in everyday practice, such as in converting the linear dimensions of the page of the book from inches to centimetres, the volume of petrol from gallons to litres, temperature between °C and °F, etc. Such re-scaling amounts to a *translation* in the terms of Figure 2.3: no information is lost in the conversion, the relationship between the old and new scale is transitive, symmetric and reflexive.

The literature, and measurement practice, offers examples of many other scale types and scaling transformations, in addition to those listed in Figure 3.9. Amplitude is often measured on a logarithmic scale; re-scaling between Cartesian, polar, elliptical or other co-ordinate systems is used when defining the location of a point in space, etc. For a more detailed description of scales and scaling, see e.g.[4], and for a comprehensive treatment consult [5].

SCALE TYPE	ADMISSIBLE SCALING TRANSFORMATIONS	COMMENTS/ EXPLANATION	DEFINING RELATIONS	APPLICATIONS
Nominal	$M' = f(M)$	f is a one-to-one mapping	* Equivalence	classification
Ordinal	$M' = f(M)$; if $M(A1) \geq M(A2)$ then $M'(A1) \geq M'(A2)$	f is monotonically increasing	* Equivalence * Greater than/Smaller than	ordering, sorting, grading
Interval	$M' = aM + b$ $a > 0$	Positive linear transformation; numerical measurement of quantities relative to arbitrary origin	* Equivalence * Greater than/Smaller than * Relative scale values	e.g. relative velocity, elapsed time, temperature (in scales other than 0K)
Ratio	$M' = aM$ $a > 0$	Similarity transformation; numerical measurement with respect to a set origin	* Equivalence * Greater than / Smaller than * Relative scale values * Ratio between scale values	e.g. distance calendar time, absolute temperature, IQ scores
Absolute	$M' = M$	The scale is not transformable: new and old scales are identical	* Equivalence * Greater than / Smaller than * Relative scale values * Ratio between scale values * Absolute scale values	counting

Figure 3.9: Main scale types of measurement

3.4 Measurability

Lord Kelvin is the source of the most frequently used quote in the history of measurement: *"What is not measurable, make measurable."* But what does measurability *mean*?

To gain insight into measurability, we try to enumerate the causes of failure to measure, and distil from these the criteria of successful measurement. To do this, we draw on the key notions of measurement, developed in this chapter.

- Measurement will fail unless the observed attribute is fully understood, and a well-defined *concept* corresponds to the attribute of interest.

   The garaging example relied on the well-established notion of 'width', applicable to all spaces and all solid objects, including motorcars. The concept (linear distance) which underlies the attribute of width is broader than is required by this specific attribute; it also supports other attributes of physical objects, such as height, depth, length, etc.

   Contrast this with the attribute 'user-friendliness'. Its measurability as a software quality attribute relies on the skill of the software community to define the concept with clarity and precision. Whether or not the attribute will be meaningful in a broader context, such as for designating a feature of the quality of manufactured objects, services, racehorses, etc., will be determined by the *power* of the concept: the domain over which it has applicability.

   Note here that 'well-defined' is a technical term, designating absence of vagueness of concept. The definition must explain clearly the nature of the attribute, leaving no doubt about its applicability to a given class of entity. Three types of definition may be distinguished [6]:

   * 'lexical' definitions, to associate the attribute with an established concept; established lexical definitions were assumed for notions such as 'width', 'area', etc., used in our earlier examples;

   * 'stipulative' definitions, to designate the attribute as a new or evolving concept, building it up from its lexically defined constituents; this was the case when defining the 'wear/tear/ environmental damage' (WTE) attribute in our 'traffic example', and this type of definition is needed in defining an emerging concept, such as 'user-friendliness';

   * 'precising' definitions, to refine concepts and eliminate border-line cases in lexical or stipulative definitions;one would expect the imposition of the WTE charge to show up weaknesses and anomalies leading to the precising definition of the attribute.

- Measurement will fail unless there exists a *model* of proven quality comprising properly defined *property variables* which represent the concept and correspond to the attribute under measurement.

  Each attribute, whether qualitative or quantitative, simple (such as 'width') or complicated (such as 'area' and WTE in the examples of this chapter), must be modelled by a property variable. How else could measurement fulfil its role of characterizing a referent by giving a value to its property variable? The variable definition must specify the type and domain of the variable. When the measurement problem calls for quantitative measures, the definition of the property variable must also include the dimension implicit in the concept.

- Measurement will fail unless there is an appropriate *symbol system* and a suitably defined scale of measurement, such that the measures satisfy the formally defined rules of measurement theory.

- Measurement will fail unless there is a suitable *practical procedure* for measuring the property, directly or indirectly. Practical considerations include the availability, at appropriate accuracy and feasible cost, of suitable equipment and personnel for taking and recording observations.

  If the property is indirectly measured then practicality of the measurement procedure will require:

  * existence of a valid *theory* for modelling the indirectly measured property, and deriving the measure from a set of directly measured auxiliary properties;

  * *practicality* of the direct measurement of a complete set of auxiliary property variables.

- Measurement will also fail unless the measures are relevant to the queries arising from the original problem, and are suited to the needs of the intended user. The design of measures must take account of *human factors*, such as consistency with the prospective users' intuitive notions, their knowledge of related attributes, their familiarity with the symbols used to designate the measures, etc.

  In relatively novice user communities confusion can arise from the use of symbols of foreign alphabets; undue significance may be attributed to crude measures encoded in numerical symbols with many places of accuracy; misleading conclusions may be drawn from quantitative symbols used to designate qualitative measures; colour codes will be mistaken by colour-blind users; etc.

Summarizing the criteria of measurability

The considerations of measurability may now be summarized as follows:

Definitional consideration.

The empirically observed attribute must be identified with an already defined well-established concept, or should be defined as a composition of well-defined concepts.

Modelling and representational consideration.

Each attribute must be modelled by a well-defined property variable, and variable values represented in a suitable symbol system.

A valid theory must be available for modelling each indirectly measured property as a structure of properties for which a procedure of direct measurement already exists.

Practical consideration of direct measurement.

A feasible procedure must exist, together with appropriate equipment, skills and other resources, for observing and recording measures of the property.

Quality considerations.

The measure must be relevant to the requirements of the original problem, must be consistent with 'human factors' requirements of usability, and obey the formal requirements of measurement theory.

To meet these criteria, it is not sufficient for the measurer to rely on common sense and local knowledge. The measurer must have access to accumulated knowledge and experience distilled into established concepts, related attribute and property definitions, validated models and theories, tried and tested procedures, and corresponding instruments and skills of measurement. This must be built on an 'infrastructure' of agreed standards and maintained units of measurement, a network of calibration services, and laws for the enforcement of standards, to be called upon if required. The standard term for the whole domain of science and technology specifically devoted to measurement is 'metrology' [7]. We adopt the term here, although it is infrequently used in this broad context, even by scientists and technologists.

3.5 Metrology

The mature disciplines of pure and applied science have evolved and agreed upon the universal system of concepts and measures (SI, [8]) of the physical universe. By assuring the measurability of basic attributes of physical objects and phenomena, metrology provides support for the whole international scientific and technological community, as well as serving the interests of the community at large.

Each major branch of science and technology builds on this general and universal metrological foundation. Each discipline complements the repertoire of metrology with its own specialist concepts and measures, and implements the measurability of domain-specific attributes through its own, specialist measurement technologies. A combination of general and specialist measurement knowledge underpins professional activity in each field, and supports the development of the individual specialist discipline. Specialization does not mean isolation and chaotic divergence of disciplines: the disciplines have a common root in the universal measurement system, and they share reliance on it. The development of each discipline creates new measurement demands, and interdisciplinary collaboration is needed to assure the development of metrology itself. Advance is achieved in all fields, through mutual support.

Young, newly created technologies have not had time to establish their roots in general science. They are striving to firm up their scientific foundations, and have identified the need for measurement to assure effective development. They attempt to compile a catalogue of well-defined specialist concepts, attributes and measurable properties, but their measurement requirements are not well formulated, and even if they were, metrology in its present state would be unprepared to meet them fully. Although they are not yet locked into the interdisciplinary metrological community, they could already derive valuable support from general metrology, and would benefit greatly from the accumulated measurement knowledge and experience of the mature sciences and technologies.

As we said, the book aims is to show how systems, models and measurement may forge links between the mature domains of science and technology and the new disciplines, such as quality management of software and systems engineering. Towards this aim, in this section we outline the structure of metrology, and illustrate the way in which the measures of attributes of specialist interest are derived from general measures of common concern.

Building on the ideas put forward by Fiok (9), one may visualize metrology classified in two ways: according to the *aspect* covered, and according to the *sphere of concern*. We say that the two must be brought together into a coherent whole by a systematic approach and *model-based measurement*.

Aspects of metrology

We distinguish four 'aspects' of metrology, listed here starting from the most abstract and ending on the most practical.

Theoretical metrology:
- the philosophy of observation and conceptualization;
- the general theory of experimentation;

- the formation of definitions;
- the formation of hypotheses, models and theories;
- scale formation, and interpretation of data as evidence;
- the theory of errors.

Methodology of metrology:
- the analysis of requirements and the design of a specific measurement scheme;
- the planning, organization and execution of measurement;
- the recording and the interpretation of results.

Technology of metrology:
- the specification, design and implementation of measuring instruments, measurement tools and measurement systems for sensing, processing, storing and displaying measurement data;
- the formulation and implementation of measurement procedures.

Legal metrology:
- The system of codes, practices, regulations and laws which concern the development, use and maintenance of the units and standards of measurement.

Spheres of concern of metrology

The 'spheres of concern' of metrology are listed here in order from the most specific to the most general.

The metrology of direct measurement of a single attribute:
- Section 3.2.1 of this chapter covers much of the theoretical aspects, and also touches on methodology.

The metrology of indirect measurement of a single attribute:
- The background is the subject matter of Section 3.2.2.

Object-oriented metrology:
- The main ideas are outlined in Section 3.2.3.

The metrology of utility:
- The principles are given in Section 3.2.4.

Subject domain oriented metrology:
- The characterization of entities within a specific domain of science or technology, including the creation of standards for the domain. Taking the domain of medicine as an example, it is the role of pathology to study disease, and characterize its manifestations by defining measurable attributes.

General metrology:

- The study of the way in which a system of basic attributes may be defined and universal standards created, such that, with their aid, entities of different specialist subject domains may be characterized in a united, general measurement scheme.

The international system of units

General metrology can claim credit for creating, and establishing as international standard, a comprehensive system of units which have proven necessary and sufficient for measuring the attributes of the physical universe. SI, the system already mentioned, has a set of 'base units'. Subject domain oriented metrology then relies on the theories and models of the physical sciences to derive, from these units, measures for all other properties required for characterizing referents of these sciences and their technologies.

Figure 3.10 gives an outline of the 'hierarchy' of SI units, showing in the left-hand column the name and dimension of the seven base units of general metrology. These are used in all disciplines, and in the construction of other measures. Convenience has dictated the inclusion of a small number of supplementary base units, such as the radian, a scalar measure of plane angle. Progressing towards the right, the figure displays derived units for a few important properties, used in many disciplines. The lines across the figure show the connection between base and derived units. The rightmost column of the figure is a selection of some of the derived units of electrical engineering. Measures of other derived attributes, such as power factor, coefficient of thermal expansion or dielectric constant, will require further compositions of measures, and corresponding progression towards the right of the figure.

The hierarchy of relations among the units form a directed graph. The source nodes of the graph are the base units. The graph is acyclic, indicating that the units of measurement for all attributes are derivable from the base units.

It would be possible, although very tedious, to construct a figure like that of Figure 3.10 for subject domain oriented metrology in any specialist field. The figure would show the way in which general metrology provides the foundations for all measures. The structure of units is implicit in general science, and in the theories of each mature scientific and technological discipline on which specialist practitioners rely in course of their daily routine, including measurement, dimensional analysis, and research into novel phenomena.

The elegant frugality and universal applicability of SI, and the rigour of the theories and models on which it rests, have been key factors in facilitating the technological development of modern societies. SI would now require

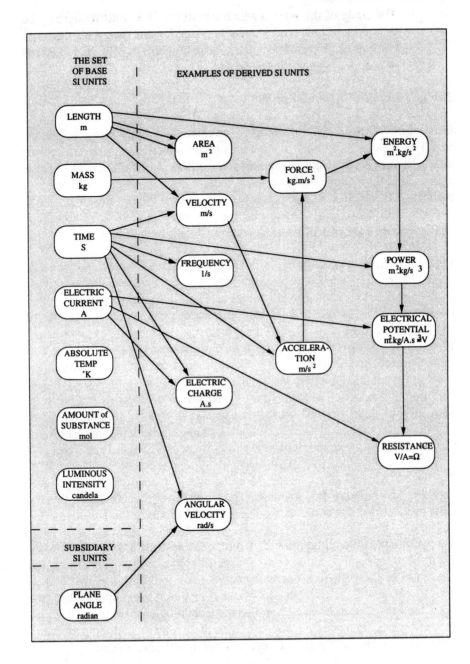

Figure 3.10: Outline of the hierarchy of SI units: the base units and samples from mechanics and electrical engineering

revision and possible extension, so that it could meet the measurement requirements of important developing domains, such as the social sciences, environmental science, computer science and their technologies. Attributes of entities within the scope of these disciplines will have to become measurable within an integrated, comprehensive system of units. This requirement may give rise to the definition of further supplementary measures, such as the bit as the unit of information, and some suitable currency unit, such as the ECU, the £ or the $.

3.6 Examples

3.6.1 Software quality measurement

The literature, standards and guidelines of software engineering offer various 'frameworks', 'models' and 'metrics methodologies' in support of software quality. To demonstrate their approach to quality measurement, we show two examples: an earlier version DP 9126 (see [10], illustrated in Figure 3.11a), and the new IEEE standard 1061 ([11]). For a comprehensive recent critique on the current ISO/IEC 9126, see e.g.([12]).

EXAMPLE: IEEE 1061: Software Quality Metrics Methodology

The standard states:

> "The use of software metrics reduces subjectivity in the assessment of software quality by providing a quantitative basis for making decisions about software quality."

The standard stipulates a "software quality metrics framework", given in Figure 3.11b. The supporting definitions discriminate:

   (i)   quality *factors*, which are "management oriented" attributes,
   (ii)   *quality subfactors*, which are "decompositions" of quality factors into their "technical components", and
   (iii) *software quality metrics*, which are "functions" producing a "single numerical value that can be interpreted as the degree to which software possesses a given attribute that affects its quality".

The standard emphasises its own "flexibility", but in its ANNEX it gives "illustrative usable definitions". It also offers "examples of factors, subfactors and metrics, and their relationships", with a four-level breakdown (into sub-subfactors?) in case of 'MAINTAINABILITY'. The relationships give hierarchical dependency, shown here diagrammatically in figure 3.11c, but do not define the structural relations for deriving the higher-order entities.

END OF IEEE 1061 EXAMPLE

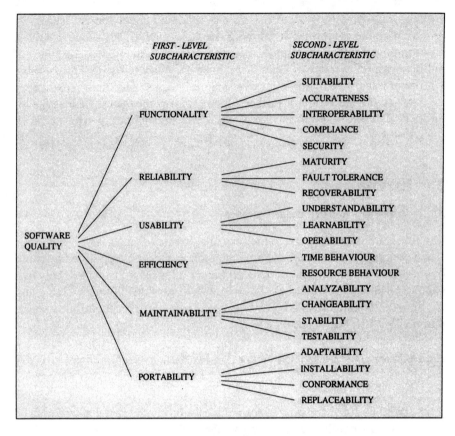

Figure 3.11a:
The composition of software quality, as given in the DP 9126 draft standard

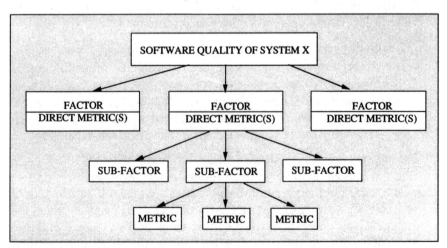

Figure 3.11b: The IEEE 1061 'Software Quality Metrics Framework'

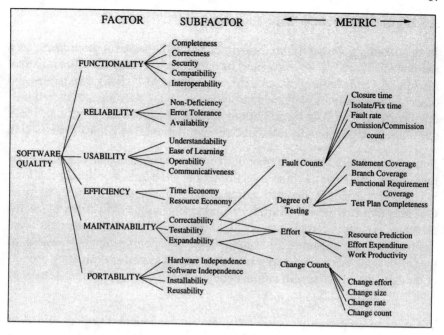

Figure 3.11c: The relationship between Software Quality, its Factors, Subfactors and some of its Metrics, suggested in IEEE 1061

At first glance, the hierarchies of figures 3.11a and 3.11c may show some likeness to the structure of SI measures, but the resemblance is superficial. In the case of the software standard, there are no universal concepts and corresponding base units, and the stipulatively defined attributes do not meet the definitional criteria. There is no measurably defined property variable set for the foundational 'metrics'. Direct measures are not identified, and there are no explicit models for deriving from them indirect measures, or to pass up the hierarchy the various multilevel 'factors', 'subfactors and 'metrics'. Although many of the attributes rely implicitly on general metrology, there are no metrological roots of software measurement for the proposed "software quality metrics methodology". In other words, a *model-based measurement scheme for software quality* is yet to be established.

3.6.2 A measurement scheme for trees

In this example we apply the principles of model-based measurement to a mathematical object, the tree. Trees have emerged as important fundamental models of a wide variety of hierarchically organized systems: companies and other social structures, service distribution systems, documentation systems, engineered assemblies, specifications, etc. Because of their clarity, they are excellent 'compositionally invariant' representations of complex systems [13].

Finite trees – the only class of trees of interest – manifest objective and directly measurable properties. From these, further objective measures emerge, allowing an object-oriented characterization of any finite member of the set of trees. One may then use these measures for composing (subjective) utility measures relevant to specific referents, such as judging the 'reasonableness' and 'uniformity' of managerial loads in the company, appraising the 'fairness' of the service provided to subscribers of distribution systems, evaluating aspects of the 'quality' of the specification, or determining the 'type' of real-life entity which the specification represents.

Trees, or more properly rooted trees, are mathematical objects, defined as follows:

Definitions

A rooted tree is a directed graph, with a unique node having no in-arc, the *root node,* a set of nodes with no out-arcs, the *leaf nodes,* all other nodes having one in-arc and one or more out-arcs, and a direct route from the root node to all other nodes.

A path of the tree is a directed route from the root node to a leaf node.

End of definitions

Figure 3.12 illustrates a variety of trees.

The development of the measurement scheme of direct, indirect, object-oriented and value measurement of trees follows the multi-layer structure of Figure 3.8.

The problem

The problem is perceived as one of producing a measurement scheme for the object-oriented characterization of any referent which is modelled as a tree, with the view to extending the scheme to value measurement. Attributes of interest include size and shape of the tree.

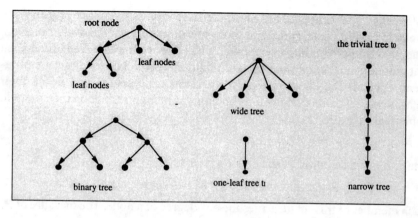

Figure 3.12: Example trees

The model

The tree displays a set of objectively observable properties and relations between these properties, in the form of a black-box model:

$$T: (\textbf{Prop}, \textbf{R}_{\text{Prop}}),$$

where   **Prop** designates the property set of trees,
and $\textbf{R}_{\text{Prop}}$ is the set of relations between properties, including the individuality relation which links all properties to the same tree.

The measures

All tree measures, whether directly or indirectly measured, are quantitative and scalar.

DIRECT AND INDIRECT MEASURES

As we have seen earlier, the identification of an attribute as directly or indirectly measurable is seldom clear-cut. This is especially true of those mathematical objects where invertable relationships exist between property variables, such that each is derivable from a set of the others. For example, in the case of trees, the attribute 'size' might be modelled in one of two property variables, either of which could be designated for direct measurement and the other for indirect, using the relation:

$$\text{node\_count} = \text{arc\_count} + 1.$$

Rather than selecting either of these, we have chosen two other properties to be classed as direct measures, and show how others may be derived from them. If another property of interest emerges which is not derivable from these, then that property must be considered as an additional candidate for direct measurement. Both of the chosen property variables are vectors, and hence their measures will be sets of numerals, holding far more information than a single numerical measure.

DIRECTLY MEASURABLE PROPERTIES.

The two proposed direct measures on trees are:

P_1:   **dispersion set** – the set of out-degrees of all nodes.

Each member of this set measures the out-degree of a node. The measure of P_1 yields a set of natural numbers.

P_2:   **path length set** – the set of all path lengths.

Each member p_i in the set P_2 is a path length. The measure of P_2 yields a set of natural numbers.

These property variables arise directly from the definition of the tree as a mathematical object. They are not independent: a change in any element of the dispersion set will cause a change in the path length set. Given these two properties, others follow as a consequence of graph theory.

DERIVED / INDIRECTLY MEASURABLE PROPERTIES

(i) Absolute measures

P_3:   **node_count** $= C(P_1)$.

The number of nodes is equal to the cardinality of the dispersion set.

P_4:   **arc_count** $= C(P_3) - 1 =$ sum of elements of P_1.

P_5:   **leaf_count** $= C(P_2)$

The number of leaves is equal to the cardinality of the path set.

P_6:   **longest_path_length** $= \max(P_2)$

P_7:   **maximum fan_out** $= \max(P_1)$

P_8:   **cover** $=$ sum of elements of P_2

(ii) *Ratio measures*

Some of the above, and some additional properties, can conveniently be defined as ratio measures, such as:

P_9: **magnitude**

The property is taken as being linearly related to node-count. The magnitude of the trivial tree t_0 is taken as the standard unit, and magnitude of any tree t_i is measured as its integer multiple:

$$P_9(t_i) = \frac{k_1 \times node\_count(t_i)}{k_1 \times node\_count(t_0)}$$

$$= \frac{P_1(t_i)}{P_1(t_0)}$$

$$= magnitude(t_i)$$

P_{10}: depth

The property is defined as linearly related to longest_path P_6. The trivial tree has no depth, and the standard unit is taken as the integer multiple of the depth of the unique 2-node tree.

$$P_{10}(t_i) = \frac{k_2 \times longest\_path(t_i)}{k_2 \times longest\_path(2\_node\ tree)}$$

$$= \frac{P_6(t_i)}{P_6(t_1)}$$

$$= depth(t_i)$$

P_{11}: width

Defined as linearly related to leaf_count. The width of the 2-node tree is taken as the standard unit of measurement, and the width of any tree t_i is its integer multiple.

$$P_{11}(t_i) = \frac{k_3 \times leaf\_count(t_i)}{k_3 \times leaf\_count(2\_node\ tree)}$$

$$= \frac{P_5(t_i)}{P_5(t_1)}$$

$$= width(t_i)$$

(iii) *Some further measures*

Measures for further properties are defined here to illustrate a way in which one may characterize the 'shape' of a tree.

P_{12}: leafiness

A proportion of the number of leaf nodes to the total number of nodes. This ratio measure has no meaning for the trivial tree t_0 (Figure 2), and for others it is a real number in the range $0 < M(P_{11}) < 1$.

$$P_{12}(t_i) = \frac{leaf\_count(t_i)}{node\_count(t_i)}$$

$$= \frac{P_5(t_i)}{P_3(t_i)}$$

$$= leafiness(t_i)$$

P_{13}: maximum_cover

This is the characteristic of an imaginary symmetrical tree which is as deep as the tree t_i under measurement, has as many leaves as t_i, and all leaves are at the longest_path_length.

$$P_{13}(t_i) = \text{longest\_path\_length}(t_i) \times \text{leaf\_count}(t_i)$$

$$= P_6(t_i) \times P_5(t_i)$$

$$= \text{maximum\_cover}(t_i)$$

P_{14}: relative_cover

This ratio measure is derived as the distribution of leaves over the tree: the deeper the leaves lie, the greater the value of relative-cover.

$$P_{14}(t_i) = \frac{\text{cover}(t_i)}{\text{maximum\_cover}(t_i)}$$

$$= \frac{\text{cover}(t_i)}{\text{longest\_path\_length}(t_i) \times \text{leaf\_count}(t_i)}$$

$$= \frac{P_8(t_i)}{P_6(t_i) \times P_5(t_i)}$$

$$= \text{relative\_cover}(t_i)$$

The magnitude of this ratio measure is a real where $0 < P_{14}(t_1) \leq 1$. For the trivial tree the measure is undefined.

P_{15}: path_ratio.

This is a ratio measure for the characteristic of the lying of the leaves, derived from P_3. It is measured as the ratio of the longest to the shortest path. $M(P_{15}) \geq 1$.

$$P_{15}(t_i) = \frac{\text{max\_path\_length}(t_i)}{\text{min\_path\_length}(t_i)}$$

$$= \frac{P_6(t_i)}{\min(P_2(t_i))}$$

$$= \text{path\_ratio}(t_i).$$

Some examples of uses

The various measures relating to the attribute of size, including counts of arcs, nodes and the magnitude measure itself, are used as crude indicators of cost, effort, resource requirements, etc. associated with referents modelled as trees.

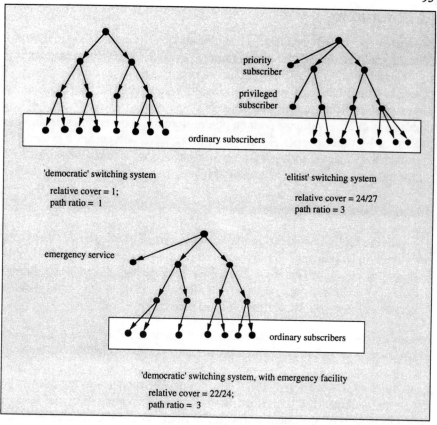

Figure 3.13: Tree models of different switching systems which share many properties, such as node_count, arc_count, magnitude, depth, etc.

Such measures would also find use, for example, in assessing system testability, and in estimating the vulnerability of systems to design error. In networks, path length is indicative of delay. Dispersion measures are useful in estimating loads on distribution nodes. Figure 3.13 shows a possible use for some of the shape measures. A further example arises in later chapters of the book.

An enquiry into the discriminatory nature of the proposed measures was conducted ([14]) on the complete set of 7-node rooted trees, 48 in number. The study showed that three properties, (leaf_count, max_path_length and cover) were sufficient to partition the 48 members of the set of 7-node trees into 37 groups, comprising 1 triple, 9 pairs, and the rest unique.

3.7 Summary

Measurement records empirical observations on a referent of the real world. Measurement is not the casual assignment of numbers to things; instead, it is the characterization of referent explicitly, with the aid of a suitable system of symbols which have meaning for the appropriate sector of the community. One must guard the integrity of measures, and assure the consistency of any measure with observations taken on other referents, at other times, in another places, or by other people. To this end, this chapter advocates a systematic approach to measurement, making explicit the reliance of measures on models, and setting out the criteria of measurability.

The model-based measurement process

The *referent* under investigation may be any type of entity, and the attributes of interest may also vary greatly: they may relate to objectively observable features, or to views and opinions. They may be simple well-understood notions, captured in a single property, or they may relate to new, complicated concepts, calling for a whole structure of properties.

Measurement is a two-step process, from referent to model, and from model to measure. A *model* is an abstraction which represents selected attributes of the referent, providing their context and meaning. The chosen attributes must be defined in property variables of the model. The model also represents the inter-relationships of the attributes, thus allowing derivation of properties and characterization of attributes which are difficult to observe directly. It is the role of the model to assure that each attribute is resolved into well-defined, formally related, directly measurable properties.

When measurement is performed, the observations are recorded as *measures:* values of the model's property variables. Criteria of measurability extend to definitional, modelling, practical and quality considerations, the latter taking account of a measure's fitness for the intended purpose and the target user community. Measures may be either qualitative or quantitative. Quantitative measures have both magnitude and dimension, the latter permitting the tracing of attributes of the referent on a universal system of units of measurement.

The model-based measurement scheme

The need for measurement may arise from a general quest for knowledge. More usually, measurement serves some decision problem, when the decision maker has to judge the referent on the basis of key properties, and the properties of interest may or may not be directly measurable. A *model-based measurement scheme* allows the deduction of each property of interest from those directly measurable; it then permits the 'object-oriented' characterization of the referent as a coherent assembly of (directly and indirectly measured) properties; finally, it explicitly states the value judgements of the decision maker in 'utility measures'.

The models of the measurement scheme lead naturally to the definition of measures for the attributes represented in the model. The direct measures of the measurement scheme form a 'knowledge base', from which indirect and utility measures are deduced by means of the model. The model may also be used inductively, revealing new properties, and pointing to new attributes of the referent which may have gone undetected or unobserved.

In practice, measurement schemes must evolve, to accommodate developments and changing requirements. To this end, the measurement scheme may be extended, modifying existing models, constructing models for new properties, and, if need be, widening the knowledge base of directly measured properties.

Metrology for model-based measurement

Metrology is the body of knowledge related to the philosophical foundations, theories, technologies and practices, and legal and administrative structures of measurement. Metrology plays a key role in defining standards and providing a comprehensive and universal system of base units to which all quantitative units of measurement are traceable. For measurement of the properties of the physical universe, this role is fulfilled by the SI units of classical metrology, supported by the models of the natural sciences. Metrology must now be extended and developed, to articulate the procedures of model-based measurement already applied in the classical domains of science and technology, and establish similarly firm foundations for model-based measurement of software, computer-based systems, complex technological systems, and referents of the social sciences.

3.8 References

1   Fenton N E (1991): "Software metrics". Chapman and Hall.

2   Kaposi A A (1990): "Classification". Task 4, BTRL Project SE00361: "Framework for Network Modelling".

3   Kyburg H E (1984): "Theory and measurement". Cambridge University Press.

4   Kaposi A A (1991): Measurement theory. Chapter 12 in "The Software Engineer's Handbook", edited by J McDermid, Butterworth.

5   Roberts F S (1979): "Measurement theory, with applications to decision-making, utility and the social sciences". Addison Wesley.

6   Copi I M (1982): "Logic", Chapter 4. Collier Macmillan.

7   BS 5233: 1986.

8   Bureau Internat des Mesures (1991): "SI - The international system of units". 6th
    edition.

9   Fiok A J, Jaworski J M, Morawski R Z, Oledzki J S, Urban A C (1988): "Theory of
    measurement in teaching metrology in engineering faculties". Measurement, Vol 6,
    No 2, pp63-68.

10 DP9126: "Software Product Evaluation - Quality Characteristics and Guidelines for
    their Use". Working Document for DIS, Work Item 97.07.13.01, JTC/SC7/WG3,
    January 5, 1990.

11 IEEE Std 1061- 1992 "IEEE Standard for a Software Quality Metrics Methodology".

12 Mellor P (1992): "Critique of ISO/IEC 9126. ISO/IEC JTC1/SC7, Doc. 92/76489.

13 Kaposi A A, Pyle I (1993): Systems are not only software. Software Engineering
    Journal, Vol. 8, January, pp 31-39. IEE.

14 Myers M (1989): "Quality assurance of specification and design of software". Ph.D
    dissertation, 1989, South Bank University.

4 MODELS OF SPECIFICATION AND DESIGN

4.1 Introduction

In daily life, the terms 'specification' and 'design' are intuitively understood, and refer to other intuitive notions, such as 'product', 'process' and 'artefact'. Through practice, these terms have acquired more precise meaning in the classical disciplines of engineering, but their casual usage still causes much uncertainty and confusion in many fields. Special concern arises in software-based systems which are increasingly entrusted with essential tasks in industry, finance, public administration and control of safety-sensitive plant. It is proving hard to specify and design such systems adequately, and to define for them standards which could assure their quality and safety.

This chapter seeks to ease these problems by offering a systematic approach to the definition of the key concepts of specification and design, as well as the entities to which they refer. To achieve this, we rely on the systems concepts introduced in Chapter 2, and the principles of measurement discussed in Chapter 3. Using these, in Section 4.2 we develop a 'product/process model' as the medium for representing the life history of artefacts, and in Section 4.3 we focus on the concepts and models of specification and design. Chapter 5 will show how they may be characterized by model-based measurement.

Why specify?

In industry, business and trade, a specification is the basis of contract between vendor and purchaser. Any deficiency in the specification can give rise to dispute, with the associated risks of adverse publicity and high legal costs. The contractual role of specifications is widely recognized, and we bear this in mind in our discussions. However, specifications have many other important uses in the life of an artefact, and we shall extend our attention to some of these.

What to specify?

A contract between vendor **X** and purchaser **Y** may relate to the supply of tangible goods, or else to abstract entities, such as software, information, and expert opinion. The contract may specify static properties of a product, or it may refer to the processes which the entity performs in operation. The subject of the contract may also be the supply of services provided by an individual, an organization, or some equipment. All of these may be the 'referent' of the contract between the parties **X** and **Y**. The contract would stipulate that **Y** pays **X** a given sum of money for the supply of the specified referent. Unless the referent, we shall call it an *artefact* (for want of a more appropriate term), regardless of whether it is a physical or abstract entity. Our prime concern in this chapter is the specification and design of artefacts, but we go on to discuss the specification and design of processes.

4.2 Products, processes, and the life history process

4.2.1 The product/process model

'Product' and 'process' are basic concepts on which we build our definition of 'artefact'. It is surprisingly hard to find any definition for the terms 'product' and 'process' in the technical literature, and where definitions are provided, they tend to be circular or not very informative. We must therefore resort to stipulating the following pair of definitions:

Definition
A product $\mathbf{P}(t)$ is a system, represented by its black-box model. The model carries the property values and 'time signature' of the product: the real time instant t, which indicates the *unique time of existence* of the product $\mathbf{P}(t)$.
End of definition

Definition
A process $\mathbf{q}(\delta t)$ is an activity of finite duration which transforms some product $\mathbf{P}_i(t_i)$ into another product $\mathbf{P}_j(t_j)$. The duration of the activity is shown as $t_j - t_i = \delta t$ and is measured on a ratio scale.
End of definition

A *product/process model* traceably represents the relationship between a process and the products which are associated with it. Figures 4.1a and 4.1b give two versions of the directed graph model of a single process. In the first, a directed arc represents a product and a node (in a box) a process; in the second, the meaning of arc and box are reversed. The two representations carry the same information. The choice between the two is a matter of convenience, and both are used extensively in this book. (Unfortunately it is not uncommon to find a mixture of the two in practice in the same diagram, resulting in confusing or meaningless diagrams.) To emphasize the distinction between the

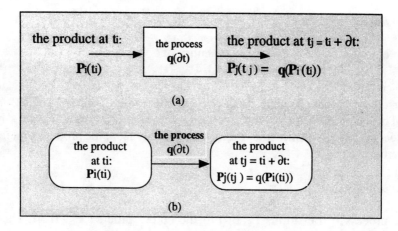

Figure 4.1: Two versions of the product/process model

instantaneous nature of products, and to contrast this with the finite time associated with processes, the time references are sometimes given explicitly in the arguments, but in practice these are frequently omitted for conciseness.

The model represents the process context-independently, as a 'closed world', showing the products which flank the process, but making no reference to the source of the incoming product and the destination of the outgoing product.

- The appearance of product P_i at t_i 'triggers' the process q.
- At t_i the process q accepts the product P_i, and at this time nothing else enters q.
- Nothing at all enters or exits the process in the time interval $\delta t = t_j - t_i$.
- The product P_j alone exits the process q at t_j, nothing else does.

To emphasize the relationship between the process and its products, the latter are frequently named $P_{in}(t_{in})$ and $P_{out}(t_{out})$. The time signature of products and the time duration of processes is sometimes omitted for conciseness, but they are always assumed.

The input product

Call the product entering the process q as $P_{in}(t_{in})$. The input product may be a single entity or a whole set $P_{in}(t_{in}) = \{ P_{in1}(t_{in}), P_{in2}(t_{in}), ... \}$, the common time signature imposing the necessary relation over the constituent products. Each product in the set may be atomic, characterized directly by its black-box model, or non-atomic, when the black-box model is derived indirectly from the product's internal structure, in the manner described in Chapter 2.

The output product

The output product may be a single entity or a whole set,

$$\mathbf{P}_{out}(t_{out}) = \{\mathbf{P}_{out1}(t_{out}), \mathbf{P}_{out2}(t_{out}),...\},$$

the common time signature imposing the synchronizing relation over the constituent output products. Each element of the output product set is generated by a parallel branch of the process. Once again, each product's black-box model may be given directly, or derived indirectly from the product's internal structure.

The process

The process $\mathbf{q}$ transforms $\mathbf{P}_{in}(t_{in})$ to the output product $\mathbf{P}_{out}(t_{out})$. This is denoted as:

$$\mathbf{P}_{out}(t_{out}) = \mathbf{q}(\mathbf{P}_{in}(t_{in})).$$

The process $\mathbf{q}$ may be atomic or non-atomic. The structure of $\mathbf{q}$ is conveniently represented as a series/parallel directed graph, built of elements of Figure 4.1a or 4.1b. Figure 4.2a is such a model of the non-atomic process of making an omelette.

The simple example illustrates several important features of processes and product/process models.

- The duration of the process $t_{out} - t_{in}$ is interrupted by events at which intermediate products emerge. The time sequence of events is ordered by the paths through the graph.

- If two or more input products enter any process, they must be synchronized. In the example, a compulsory 'wait a bit' process is imposed on the 'BeatenEggs' such that they could join 'HotButter' at the entry to the 'cook eggs' process.

- The passage of time is always associated with a process, even if there is no intention to change the product. The 'wait a bit' process intends to cause no change; rather the reverse: the success of the omelette is conditional on the invariance of key properties of 'BeatenEggs'. Nevertheless, the process is implicit in the passage of time, indicated by the different time signatures of 'BeatenEggs' on either side of 'wait a bit', and this means that property changes are possible.

- The product/process model may be refined further, if required. In the example, 'break & beat eggs' may be refined into a series/parallel process whose products are 'WholeEggs' and 'BrokenEggs'. Also, a more elaborate version may show aspects of resource and refuse management (e.g. use of energy for heating, disposal of eggshells).

- The details of the product/process model may be suppressed, if required, by structural analysis of the model. At the top level, the series/parallel process of Figure 4.2a is a sequence of two processes (q_4, and the combination of q_1, q_2, q_3); altogether, the structure may be represented by the tree of series / parallel homogenes of Figure 4.2b.

- Since time cannot flow backwards, all product/process models are acyclic directed graphs.

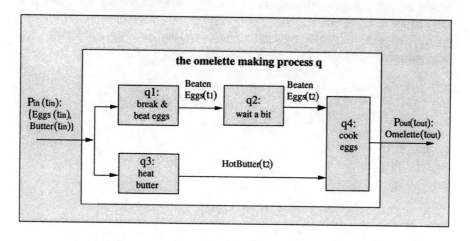

Figure 4.2a: Product/process model of making an omelette

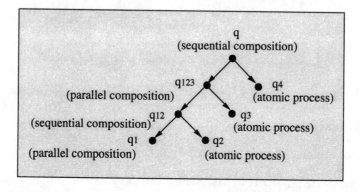

Figure 4.2b: The process structure of the product/process model of Figure 4.2a

An example of a 'generic process model' from the literature

Figure 4.3a is drawn from one of the current draft standards of the civil avionics industry. Since the model shows a single process, it must start at a given time;

thus, the two constituent input products ('Inputs' and 'Feedback from Processes Using Outputs') must have the same time signature. Also, since time moves forward (and, as we said, there is a single process), the same delay is imposed on both output products ('Outputs' and 'Feedback to Processes Which Provide Inputs'). It is interesting to note that the draft standard makes no use of its own 'generic process model': none of the process diagrams apply it. Perhaps it would have been more helpful, and less confusing, to draw this figure differently, gathering together the components of the input and output products, and pointing all arrows in the same direction. Figure 4.3b tells the same story as 4.3a, but it adheres to the usual convention of representing the passage of time from left to right on the page, and is clearly consistent with the product/process models shown earlier in this chapter.

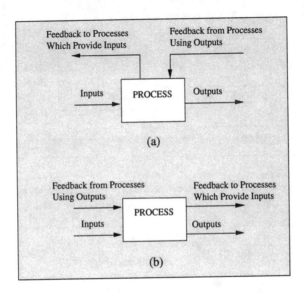

Figure 4.3: Two pictures showing the same 'generic process model'

4.2.2 Artefacts

The word 'artefact' is used to describe a broad range of entities:

- inanimate physical objects, such as a rock, an aeroplane, or an item of computing equipment;
- live objects, such as a plant, an animal, a human being, or a social system, such as a design team;
- an abstract entity, such as a specification, a design, a theory, or a computer program;
- any part of these things, such as the result of work in progress on the factory floor towards creating a marketable entity.

We assume that an artefact is 'born' and 'dies' at distinct times. At these instances, it appears as the product $P_{birth}(t_{birth})$ and $P_{death}(t_{death})$, respectively. Between the two extreme time instances of its existence, $t_{b(irth)}$ and $t_{d(eath)}$, the artefact A evolves through a *life history process,* from some notion A_0 in the mind of its initiator, to its ultimate manifestation A_n, when its life is over. (Note here that the term 'life cycle', well-established in the software world, is misleading. As we have noted before, time cannot flow backward, and all product/process models are acyclic.)

The time interval $\delta t = t_{d(eath)} - t_{b(irth)}$ is the *lifetime* of the artefact. We consider here only the discrete finite case: the artefact is modelled by a finite set of properties, is of finite lifetime, and the life history processes are described in terms of time as a discrete variable.

The life history q of the artefact A may be viewed as an indivisible 'atomic' process, or it may be described as a structure of subprocesses or *phases,* such as specification, design, implementation, operation during service life, maintenance and modification, and ultimate disposal. The phases of the life history are subprocesses $q_1 \ldots q_n$ of the complete process q. The structural relation over the subprocesses may be a sequence, or else parts of the structure may branch and merge to include parallelism, noting again that the structure can always be modelled by an acyclic graph, resolvable to a sequence of subprocesses. Each subprocess $q_i \in q$ occupies a finite elapsed time slot δt_i, within the elapsed time t of the process as a whole. Each subprocess in the life history of artefact A may itself be atomic, or it may be further resolved, possibly on several levels, into a sequential structure of its own sub-subprocesses. From our present viewpoint, the subprocesses of particular interest will be the creation of the specification and the design.

Figure 4.4 is a simplified sequential model of the life history process of some artefact A. For the time being we ignore resource considerations. The whole process, and each of its sub-processes, is flanked by a distinct product. These are the *versions* of the same artefact at various time instants, each version carrying the time signature of its black-box model. The cartoons showing source and destination are only there to decorate the picture and assist in visualizing the message; they are not part of the language of the product/process model.

We may now use the product/process model to firm up the notion of 'artefact' as a sequence of products, created by the processes of a lifetime. The processes involve many property changes, but throughout these the artefact preserves its *identity*, described as an "absolute or essential sameness" [1]: that combination of features which the artefact retains throughout its existence, and which allows it to be uniquely identified as itself.

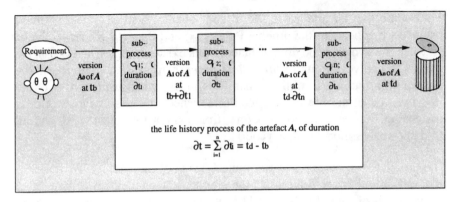

Figure 4.4: The simplified representation of the life history of the artefact A

We may define 'artefact' as follows:

Definition

An artefact A is a sequence of products $A = \{A_0, A_1, A_2, .., A_n\}$, called 'versions' of A. The black-box model of any version $A_i \in A$ contains the subset of properties $\mathbf{Prop}_A \in \mathbf{Prop}_i$, called 'the artefact signature'. The artefact signature embodies the intuitive notion of the 'identity' of the artefact: it is *invariant* throughout the artefact's lifetime, and is *unique* to the given artefact .

End of definition

As an example, consider the 'artefact' as a manufactured product, such as a computer, motorcar or measuring instrument. Its black-box model includes a serial number, which is part of the 'artefact signature'. The car may undergo major alterations during its lifetime, such as re-spraying, engine change, etc., but throughout these it maintains its serial number which uniquely designates its identity. If the 'artefact' is a person, then most physical and mental characteristics will vary during the lifetime, and even name or sex may be changed, but the individual's 'artefact signature' is maintained by means of a set of identity-carrying properties.

4.2.3 Artefact families

Consider for a moment the case where an artefact is an industrial product, one of a mass-produced batch. Each item in the batch will have its own identity and its own artefact signature, but there will be some set of attributes common to all members of the batch which distinguishes it from the members of any other batch. We call such a collection of items an *artefact family*.

Definition

An 'artefact family' F is a set of artefacts, $F=\{A_1, A_2, ..., A_n\}$,whose initial life history phases are common, but thereafter the black-box models of individuals diverge. Each family member will have a unique 'artefact signature', while throughout their lifetime the black-box model of family members continues to retain a subset of properties, called 'the family signature', common to all family members.

End of definition

The set of cars leaving a production line, all cars having been produced to the same specification and the same design, and having met the same quality assurance requirements, would form an artefact family. The life history of individuals would start with common specification and design phases, but individual behaviours will begin to diverge, each acquiring a unique identity during implementation, within limits guarded by the quality procedures of the manufacturing process. Family resemblance will be sustained through life, because of shared specification and design, but after leaving the supplier, the behaviour of individual cars will diverge rapidly, as members of the same production batch will be supplied to different dealers and sold to different owners, as each individual is put to different use and is serviced differently. Hence the history of the whole family can only be described by the history of a 'typical' individual, or by a generic framework which maps out the 'admissible' courses of life-history processes of family members under normal conditions.

The specification and design problems of complete artefact families is one of the important topics beyond the scope of this book; here we concentrate on specification and design, during which there is no difference between family members, and the whole artefact family can be regarded as though it were an individual artefact.

4.2.4 Processes in practice

Associated with the version set $A = \{A_0, A_1, A_2, ...\}$ of artefact A is the sequence of 'phase' processes $q = (q_1, q_2, ..., q_n)$, as we have seen in Figure 4.4. Any process $q_i \in q$ acts on version A_{i-1} to generate version A_i.

Consider now a phase, or a sequence of phases, in the life history process of artefact A, and assume that the process q transforms some version $A_i(t_i)$ to a new version $A_j(t_j)$. Experience tells us that no real-life process can be accomplished without expending some resources, not even one which deals with entirely abstract artefacts, such as computer programs, or mathematical theorems. In practice, resources will have to be available at the start of the process at time t_i. Part of these resources will be absorbed to add value to the product A_i; the remainder will be accounted for at the end of the process at time t_j, as recoverable by-products, or non-recoverable waste-products and other expenditure.

We denote the resources at the input and output of the process as $B_i(t_i)$ and $B_j(t_j)$, respectively. Figure 4.5 shows the process q, acting on the composite product $P_i(t_i)$, made up of a version A_i of the required artefact A, and B_i: the necessary input resources. The process generates the composite product $P_j(t_j)$, made up of versions A_j of A, and the resource output B_j. Only in the ideal case can a process be considered 'lossless'. A lossless process q' would be an abstraction: a *model* of the realistic process q, neglecting the resources utilized and generated by q.

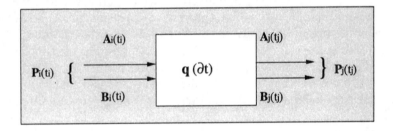

Figure 4.5: The process q, acting on the composite product P_i, and generating the composite product P_j

Recall now that P_i, P_j and q can be considered to form a 'closed world'. This means that, for the duration of q, the laws of conservation of energy apply to the physical manifestations of P_i and P_j. The process q may act as an energy converter, but not as a dissipater or supplier of energy. Similarly, if q transforms information from one type of representation to another, it can neither lose any of it permanently, nor can it generate any new information. Thus, the energy and information content of P_i and P_j at the input and output of process q is the same. The definition of these invariants opens the way to the characterization of the resource management of processes – an important topic, but again one beyond the scope of this book.

4.2.5 A more realistic model of the life history process of an artefact

We now revisit Figure 4.4's description of the life history process.

- In the ideal case, where the process can be regarded as lossless, the life history will be as shown in Figure 4.4.

- In practical cases, where the life history process absorbs an amount of finite resources, the process will need access to a 'pool' from which it draws these resources, and to which all spent resources are returned (Figure 4.6). The 'resource management process' runs alongside the artefact'slife history process.

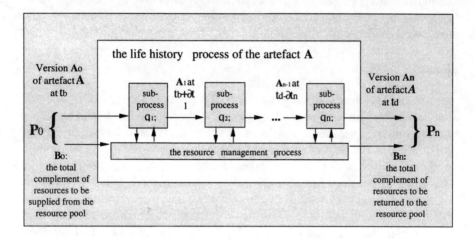

Figure 4.6: The life history process of the artefact A, with its associated resource management process

We may now define the life history process as follows:

Definition

The life history process **q** of the artefact A has duration $\partial t = t_{d(eath)} - t_{b(irth)}$, which is the *lifetime* of A. The process **q** acts on the product $\mathbf{P}_0$ and generates the product $\mathbf{P}_n$. $\mathbf{P}_0$ is the composite input product of the process, consisting of the initial version $\mathbf{A}_0$ of the artefact A, together with the entire complement of requisite resources, $\mathbf{B}_0$. $\mathbf{P}_n$ is the composite output product, consisting of the 'dead' version $\mathbf{A}_n$ of artefact A, together with the totality of resources $\mathbf{B}_n$, generated or dissipated in course of the lifetime.

End of definition

The process **q** may be atomic, or it may be constructed as a sequence of phase processes $\mathbf{q} = seq(\mathbf{q}_1, .\mathbf{q}_2.., \mathbf{q}_n)$, where any phase process $\mathbf{q}_i(t_i)$ acts on the composite product $\mathbf{P}_{i-1} = \{\mathbf{A}_{i-1} \in \mathbf{A}, \mathbf{B}_{i-1}\}$ to generate the composite product $\mathbf{P}_i = \{\mathbf{A}_i, \mathbf{B}_i\}$, where $\mathbf{A}_{i-1}, \mathbf{A}_i$ are versions of A, and $\mathbf{B}_{i-1}, \mathbf{B}_i$ are resources. Each phase process is ultimately resolvable into a sequence of atomic subprocesses.

4.2.6 Life history scenarios

Figures 4.4 and 4.6 give only schematic representations of a sequential life
history process; they do not elaborate on the kind of activity which takes place
in each process phase. We now take a brief look at the artefact's life history
from this viewpoint, and outline a few different life history scenarios. To do
this, it is not essential to take resource management into account explicitly;
thus, we can start from the simple life history schema of Figure 4.4.

Bespoke artefacts

A 'bespoke' artefact is made to the order of a client, to satisfy his/her specific
requirements. The ideal scenario is one of uninterrupted forward progression
from requirement through all phases of development and service operation,
each process accomplishing its prescribed function at the first attempt.
A typical sequence of phase processes to which an artefact is subject may be
described along the lines of Figure 4.7, although there will be variations
for different kinds of artefacts. (Time signatures have been omitted to keep the
figure simple, but note that versions of the artefact which occur more than
once under the same name, i.e. 'The Scrap', are distinguished by their
individual time signature.) An example of such a straightforward progression
is the original version of the well-known 'waterfall' model of software
development.

In practice a bespoke artefact is likely to go through several iterations of the
phases and subphases of specification, design and implementation before it is
declared ready and fit for service. In unfortunate cases the iteration may even
involve later phases, for example when an artefact is recalled from service for
modification of its design. To represent this, Boehm's refined waterfall model
[2] has become widely adopted in the software industry. The schema of a
variant of this is shown in Figure 4.8a, w_i standing for process. In this model,
arrows pointing down the 'waterfall' represent progression, and arrows
pointing upwards and backwards show regression to earlier phases so as to
remedy mistakes or deficiencies arising at the previous process.

Note that the waterfall model is not a product/process model, because it does
not obey the syntax rules. Figure 4.8b shows an instance of the 'waterfall'
having been transformed into a product/process model (all products and time
signatures omitted). Clearly, the story unfolding in the figure is far from ideal.
To reconstruct the possible causes of the regressions, assume that the functions
performed by the processes w_1 to w_5 are similar to q_1 to q_5 in Figure 4.7.
The first setback may have occurred because the initial design produced by w_2
failed to satisfy the specification. There is no point in repeating the deficient

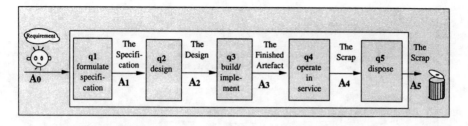

Figure 4.7: Idealized life history process of a bespoke artefact

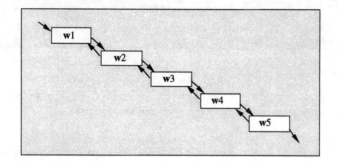

Figure 4.8a: A schematic version of the 'waterfall model'

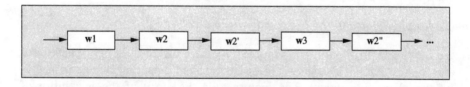

Figure 4.8b: Product/process model, showing the start of a possible progression through the 'waterfall model'

process of w_2; w_2' is a modified design process. The second regress may have been caused by the impossibility, or unfeasibility, of implementing the design in practice: the process must be repeated, with or without modification.

Other life history scenarios

Most artefacts are not bespoke: they are mass-produced, speculatively developed, evolved from previous designs, 'customer-engineered' (a generic

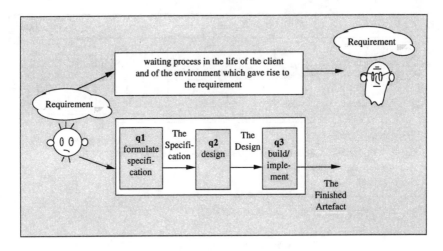

Figure 4.9:
The product/process model of the supply phase of an artefact, and the parallel
waiting process transforming the client and the requirement

design being adapted to the demands of individual customers), or will have
other, very different life histories from those just considered. Our present
interest is in the early periods of the life history, leading up to service operation.
We refer to these, collectively, as the 'supply phase'.

4.2.7 The supply phase of the life history process

The supply phase in the life of a bespoke artefact

Following from Figure 4.7, Figure 4.9 shows the ideal supply phase of a
bespoke artefact. It also shows, in the background, a parallel process in the life
of the client's company. The picture tells the story of the instability of
requirement. This is a feature of almost all projects, and a threat in all realistic
situations. Many a failed or unsatisfactorily completed project disappoints the
client, and proves to be unfit for purpose, although the artefact may comply
fully with the stated specification. If the project is complex and the supply
phase is long, the original requirement (and the personality of the client) may
have changed, and the specification will have become obsolete long before the
artefact is released into service. A well-publicized case is the development of
the M25 motorway which encircles London: by the time the construction
project was completed, the capacity for which the road was planned had proven
totally inadequate to meet the demand. In practice, the client may or may not
be aware of the changed requirement, and if there is a dispute, the supplier may
well win in the courts, but the project would still have failed, and the supplier
will have lost the confidence, or at least the goodwill, of the client.

The remedy lies in prevention: the supplier must foresee that the requirement will change during development, monitor the change, and assure, by choice of an appropriate life history model and engineering skill, that the specification and design are robust enough to withstand it.

Other supply phase scenarios

All supply activity starts with some requirement, and successful supply phases deliver a finished artefact, but the requirement is not always generated by the client. To satisfy the requirement, the phase processes (specification, design, build/implement) may have to be reordered, or reversed. We leave the reader the task of constructing product/process models for the following scenarios, all of which are quoted from actual industrial and business practice:

- The supplier company has decided to develop and market a completely novel artefact whose manufacture has become feasible as a result of a recent technological breakthrough.

- The supplier company has designed an artefact without first drawing up a specification. (The situation may have arisen because of weak management, poor documentation standards, or from a deliberate policy of speculative development in order to test what is feasible.) The company's recently imposed house standard demands that an authorized specification document should accompany the design before it can pass into the build/implement phase.

- The supplier company has just taken over a competitor, and wishes to market an existing stock of finished artefacts. At the take-over, the designers have left the company, leaving behind no usable design and specification documentation, but the industrial standard demands comprehensive documentation before the artefact can be certified as fit for the market. The further plan is to enhance the specification and update the design.

- The customer presents the supplier with an artefact which has been produced by a third party. The client's requirement is that the supplier should provide "one just like it", except for a list of changes and extra features. With the exception of a commercial leaflet and user handbook, no supporting documentation is available.

The sample scenarios just discussed call for an examination of specification and design, not as processes, but as products in the life history process.

4.3 Specification of artefacts

4.3.1 Specification and design as products

Figure 4.10 shows the first two activities of the supply phase of some artefact A. The figure is representative of both bespoke and mass-produced artefacts. Using the concepts developed in Chapter 2, the figure divides into two parts, showing that the requirement (A_0) exists in the real world (if only in the mind of the client), whereas both the specification (A_1) and the design (A_2) are entities in the world of models. q_1 is a modelling process which turns the requirement A_0 into the specification A_1. q_2 is a process which is the converse of modelling, to which we referred as 'reification' or 'instantiation': moving from the general to the particular. In this context instantiation adds detail to the specification, showing *how* it may be achieved by the design A_2, such that A_1 should be a model of A_2.

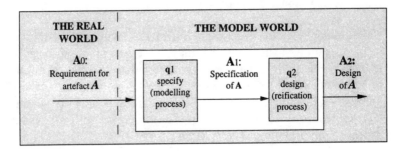

Figure 4.10: Specification and design phases of the life history process

To be valid, the specification A_1 must capture *at least* those attributes for which the client's requirement calls, and may stipulate other properties besides, such as external standards or internal company standards. The role of the specification is to express the attributes in the measurable properties of the black-box model of the future system A. The quality of the specification, as a product of the process q_1, must be ascertained before initiating the design process.

To be valid, the design A_2 must satisfy the specification A_1, but must also contain further information; otherwise the process q_2 will have achieved nothing. The extra detail comprises the structure of the artefact A, and the design is complete when it is resolved into the set of its atomic, readily available component parts of known behaviour. The validity of A_2 as the structural model of A means that it is an implicit representation of A_1, the black-box model of A, and A_1 must be derivable from A_2 in the manner discussed in Chapter 2. (Note here the need to distinguish between two interpretations of A_2. In the context of the life history process, A_2 is a structural model of artefact A. Viewed context-independently, it is a product in its own right. As such, its properties constitute the black-box model of the product as created by the process q_2.)

We may now define the two versions of artefact A, created by the specification and design processes, as follows:

Definition

Let A_0 be the requirement for some artefact A. Then the specification A_1 of A is a product which models the required attributes of A_0. The black-box model of the product A_1 takes the standard form $M_B(A_0) = ($**Prop**, $R_{Prop})$, where **Prop** is the set of property variables which, with appropriate values, capture the required attributes of A_0. R_{Prop} is the set of relations over **Prop**, including the individuality relation which carries the time signature of A_1 in the life history process of A.

End of definition

Definition

The design A_2 of artefact A, whose specification is A_1, is a structural model over atomic component parts of known specification and common time signature, such that the specification, derived from A_2, should satisfy the specification A_1.

The structural model takes the standard form $M_S(A_2) = ($**Comp**, $R_{Comp})$, where the set **Comp** comprises the specifications of the atomic components, and the relation set R_{Comp} is the structure.

End of definition

4.3.2 Modelling the specification process

When the artefact is relatively simple, the concepts of the application domain are well understood, and the requirement already implies the model of the specification which would be used by all those who are expert in the field. In such cases it is sometimes possible to capture the requirement of the future artefact directly in measurable properties. In conversation with the client, the designer of an electronic filter may write down the parameters of the frequency response, and the architect of a dwelling house of stereotyped layout may draw up the key dimensions immediately. In most situations however, the supplier will need time to analyze the requirement, and select a suitable model deliberately, before the specification can be formulated in parameters of a black-box model. As an intermediate stage between requirement and formal specification, the requirement will have to be expressed in a convenient medium, such as English text complemented by diagrams, tables or graphs, comprehensible to client and supplier. This halfway house between the unstated requirement and the formal specification is referred to as 'require-ment specification'. The term 'specification', applied to the artefact A, is reserved for the formal specification at the end of the specification stage: the product which triggers the design process.

Figure 4.11 illustrates the specification process q_1 as a sequence of subprocesses q_{11} and q_{12}. The requirement specification – the intermediate product A_{11} of the specification process – can be drafted either by the client or the supplier. As the first expression of the requirement, it is the basis of the contract, and has the authorization of both parties. The (formal) specification A_1 might give the property set of the artefact's black-box model as parameter values of a set of equations, as points in a graph, or as symbols and dimensions of diagrams. The black-box model may also be implicit, for example the representation of the artefact in an operational model, such as a working prototype.

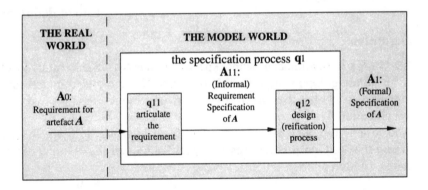

Figure 4.11: A two-step specification process.

In practice a complex system might undergo a more refined, multi-phase specification process, with intermediate products A_{11}, A_{12},... Each inter-mediate requirement specification in the sequence would tend to be more technical than the previous, and further removed from the client's sphere of intuitive understanding. Each subsequent version would express the requirement with increasing explicitness and precision. It is advisable for versions of at least the early phases to be referred back to the client before declaring the specification A_1 as valid, and releasing it to trigger the design phase. Communication of advanced versions of the requirement specification may require considerable effort from the supplier, either involving the educa-tion of the client, or, for large and costly projects, the building of real-life demonstrators, such as mock-up systems, feasibility study equipment, and executable specifications.

Methods for capturing requirements for complex systems, representations of requirements in formal specifications, and formulation of criteria for judging the quality of specifications, are major topics of current research. It is sufficient to say here that a 'good' specification is complete, consistent, concise, comprehensible, testable [3] and unbiased [4].

- Completeness means that the specification states *all* the properties of the artefact which correspond to attributes required by a user or external observer.
- Consistency means that the specification should come with assurance that the stated properties are *compatible* with each other, such that it should be possible to satisfy them simultaneously.
- Conciseness means that the specification should state *only* the relevant properties. Conciseness serves clarity, and thus protects against error.
- Comprehensibility is a practical requirement protecting the customer and assisting the supplier by assuring that the specification document is an effective medium of *communication* between the two.
- Testability also relies on conciseness for assuring that the substantial resources needed for good testing are not wasted on superfluous or secondary features. In addition, testability calls for measurability: properties should be given in a format capable of objective, *practical validation* at tolerable cost.
- Lack of bias means that the specification states only *what* the user requires, but refrains from constraining the supplier in deciding *how* the requirement should be met, keeping open the supplier's options.

These criteria are met if the specification is a 'good' black-box model of the artefact which defines the required attributes as values of measurable properties, in accord with the principles, procedures and criteria laid down in Chapters 2 and 3.

In practice, the requirement is not confined to the attributes of the new artefact, but will call for these attributes also to be sustained in service operation. One of the hardest parts of constructing the specification is to assure, with reasonable confidence, that predicted behaviour is translated into parameters measured at the end of the supply phase. When specification characterizes not just an individual bespoke artefact but a whole artefact family, then the specification is *generic,* in the sense of representing the prediction of variation over all reasonable possible life histories of family members, and setting bounds of admissible parameter variations.

Specifications often include some items of data which are, strictly speaking, superfluous. Such specifications contain *redundancy*: detail which is deducible from other parts of the specification, and can be omitted without loss of information. Casual redundancy is contrary to conciseness, and is to be avoided. However, redundancy can be put to good use, because it affords cross-checking of properties, so as to enhance confidence in their consistency with each other, and with the client's actual requirements.

4.3.3 Descriptive and prescriptive specifications

In a legal context (5), 'specification' is defined as "a detailed description of building, engineering and other works executed, or *proposed* to be executed" (our italics). The definition indicates two sorts of uses of specifications: actual and futuristic. To emphasize the difference, we refer to a specification as *descriptive* or *prescriptive*.

A descriptive specification is *factual*. It states the characteristics of a given version of an existing real-life artefact at the time of observation. In this chapter the emphasis has been on prescriptive specifications used in creating a new artefact, but we have also encountered several uses of descriptive specifications in discussion of various life history scenarios.

A prescriptive specification stipulates the *bounds of requirements* which the finished version of the artefact must satisfy when in due course the design is implemented. The descriptive specification of the system would normally differ from its prescriptive specification. The difference is acceptable to the customer, as long as the descriptive specification is *no less stringent* than the one prescribed.

The relationship between prescriptive and descriptive specifications may best be appreciated by envisaging that the **k** number of orthogonal properties of the specification of an artefact define the co-ordinates of a **k**-dimensional *design space*.

- The *prescriptive specification* sets the boundary of a *specification domain* in the design space within which the behaviour of the future artefact must reside. Consistency – a criterion of 'good' specification – assures that the design problem is solvable: the attributes of the requirement are not mutually contradictory, and the prescriptive specification domain (the 'PS domain') is a space which contains *at least* one point. In general, the specification domain is not a single point but has finite dimensions, which means that any point in the space would constitute a satisfactory design. Since the solution to the design problem is not unique, the designer has some freedom to choose *how* to meet the specification. The tighter the prescriptive specification, the smaller the specification domain, the less the freedom of the designer, and the harder to find and realize a design which will satisfy it.

- The *descriptive specification* defines a *point* in the design space. The design conforms to the prescriptive specification if the 'DS point' (the point of the descriptive specification) falls within the domain set by the prescriptive specification.

In practice, implementation is subject to uncertainties, such as finite tolerances and inevitable variations of manufacturing processes. These 'smudge' the point of the descriptive specification into an *implementation domain* of finite size around the ideal DS point. In case of a mass-produced artefact family, the DS point of the given version of some individual artefacts may fall within, others outside the specification domain. The former will pass, the latter fail the test of quality assurance. Only in the limiting case will the descriptive specification be exactly equal to the prescriptive specification, and such precise coincidence could only be detected by ideal measuring instruments of perfect accuracy. In general, the relationship between prescriptive and descriptive specification is that the point defined by the latter must fall within the domain defined by the former. This is expressed formally by the implication relation '←', which must hold between the two:

$$\text{Prescriptive specification} \leftarrow \text{Descriptive specification} \quad\text{...........Eq 4.1}$$

Assume now that some prescriptive specification S_P has been given for the two-level system in Figure 4.12, and the design returns the descriptive specification S_D. Then the design meets the prescriptive specification just as long as the relation $S_P \leftarrow S_D$ obtains. Theoretical proof or practical validation is required to ascertain that in the normal operational domain the artefact is compliant, and equation 4.1 holds.

Figure 4.12 shows a bipartite structure: the design of the two-level system, under conditions when, for each component S_i of the given version of the artefact, a prescriptive specification S_{iP} is defined, and a descriptive specification S_{iD} is actually achieved. The figure shows that the design is correct: the implication relation holds, individually, for each component, and the structure is such that, for the artefact as a whole, the requirement of equation 4.1 is satisfied.

4.3.4 Black-box models as generic specifications

As we have seen, the specification of a product $P(t)$ is defined by its black-box model which contains the set of values of its property variables. The black-box model will take the form set out in equation 2.5:

$$M_B(P(t)) = (\textbf{Prop}, \textbf{R}_{\text{Prop}}),$$

where   **Prop** is the set of **n** property variable values of $P(t)$,
and $\textbf{R}_{\text{Prop}}$ is the relation set over the property variables, including the mandatory individuality relation $\textbf{rp}_1$ which carries the time signature of the product.

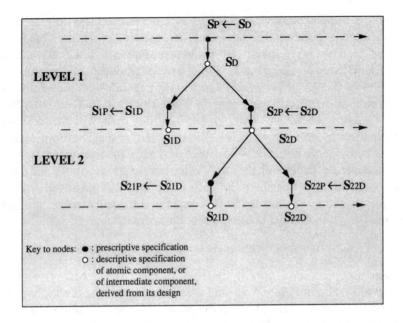

Figure 4.12: Prescriptive and descriptive specification of a two-level system

Accordingly, the black-box model defines a 'specification space' of $(n+1)$ dimensions, and the $(n+1)$ data items of the product $\mathbf{P}(t)$ will define the specific point occupied by $\mathbf{P}(t)$ in the specification space, with the time parameter and all property variables tied to specific values.

Consider now the case where the requirement is not confined to the individual product $\mathbf{P}(t)$, but calls for the specification of a whole product *class*, with $\mathbf{P}(t)$ as member of the class. The black-box model specifies a *genus* of products: the population of the complete class which fall within the type and domain of the $(n+1)$ timing and property variables of the black-box model. (The terminology taken from classical scientific philosophy, e.g. [6]). As we have just seen, an individual product in the class will be narrowed down by $(n+1)$ data items to a single point in the black-box model's specification domain. A *species* of products will be a sub-class of the product genus, with some properties defined by value, others left as variables. The specification of the species will be given by the 'size' of its domain, measurable, for example, by the number of points within the $(n+1)$-dimensional finite specification space, or by the number of the $(n+1)$ members of the black-box model's parameter set which are left as variables. We shall have occasion to define a measure for a related attribute of specifications in Chapter 5.

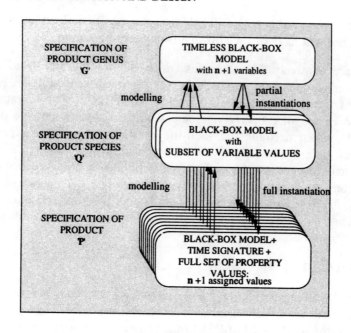

Figure 4.13:
Specification of a product, a product species and a complete product genus

Figure 4.13 illustrates three types of specification. The most general is the specification of a whole *product genus* **G**; the most detailed is the specification of the *individual product* **P**(t). Between the two is the specification of the *product species* **Q**. The figure shows the relationship between the three classes of specifications: *modelling* leads from the more specific to the more general, and adding detail by *instantiation* leads from the more general to the more specific.

Of course, specification of product species may be of many sorts, depending on the number and choice of property variables defined by value. We single out a way to subdivide the product genus **G** into species which has been found useful in several situations. Here :

$$G = \{ Q_1, Q_2, \dots \},$$

where   **G** is the genus of products, comprising individual products which have the same black-box model other than their time signature, and differ in any or all of their property values,

and $Q_i \in G$, $Q_i \neq Q_j$, is the species of products with identical property values, members of the species differing only in their time signature.

Now any product $P(t) \in Q_i$ will belong to the species defined by its property values, but will have its own distinguishing time signature.

There are relatively few situations which call for the specification of an individual product; in most practical cases generic specifications are required. All prescriptive specifications are generic, defining bounds, rather than specific values, of property variables. Generic specifications are used, for example, when defining the shelf-life of foodstuff and other perishable artefacts; here the bounds of the time signature imply tolerable limits of variation of other property variables. A generic specification is called for even where the specification is descriptive, characterizing a unique individual, because in practice measures can only be obtained with finite accuracy, and the uncertainty smudges the point-specification into a domain of finite size.

4.4 Design of artefacts

In accord with our definition, the design process converts the product speci-fication – the black-box model with corresponding data set of a product $P(t)$ – into a structural model of $P(t)$, of the form defined in equation 2.6:

$$M_S(P(t)) = (Comp, R_{Comp}),$$

> where **Comp** is the set of components of $P(t)$, defined by their black-box model and common time signature,
> and R_{Comp} is the structure which the design imposes on the component black-box model set.

The task of the design process is to select components of appropriate specification, and impose on them a suitable structure, such that the descriptive specification of the resulting product should meet the prescribed specification. *Verification* of design means ascertaining the conformance of the proposed structural model with the prescribed black-box model. *Validation* of design means obtaining experimental evidence of conformance after implementa-tion, at the end of the supply phase.

4.4.1 Modelling the design process

In all cases of practical interest, the product is too complex, and the design process too complicated, for the design to be invented and verified in a single step. This brings about the imposition of various house rules, standard 'design methodologies' and project management disciplines, whose aim is to reduce the risk of extensive re-designs by structuring the design process of stages, and verifying the product resulting from each stage before progressing to the next stage.

A SYSTEM DESIGN EXAMPLE

We give a brief outline of a product/process model, documented in handouts of one of our own industrial courses ([7]), for the supply process of a typical software system within an avionics system. To harmonize with civil avionics standards and existing practices, the model is 'vertically' composed as a 6-level structure (L0 to L5), shown in Figure 4.14a - d, with further possible vertical subdivisions within levels 3 and 4 in case of complex projects. Figure 4.14a shows the supply phase of the life history of the system, comprising the processes of specification, design and implementation. The design process within this is broken down to two parallel processes (Figure 4.14b). The computer system design is also carried out as two parallel processes: hardware system design, and software system design (Figure 4.14c). The software system design is modelled as a three-level series/parallel structure of processes, shown in Figure 4.14d, 'horizontally' composed as a sequence of 'level' processes. The set of diagrams leads to a nested product/process model of the supply part of the system's life history process. To assure the quality of the design and minimize the risk of extensive regression across the subphases of the design process, each move from one design level to the next is guarded by testing.

Figure 4.14e, reproduced directly from the course handout of ([7]), is presented here without further explanation. It shows in greater detail the internal series/parallel composition of some 'level n process' and the test procedures within it, together with the outline of the 'level n+1 process' which supplies the detailed designs of component parts. The figure leads to a product/process model of the same type as others in Figures 4.14, arrows representing products, and boxes standing for processes. The model is directly applicable to software design processes in levels 2, 3 and 4, but can be interpreted readily for design processes at other levels. In these interpretations, the figure explains the processes carried out in the blank boxes in Figures 4.14b and 4.14c.

END OF SYSTEM DESIGN EXAMPLE

The result of the design process

Figure 4.14e emphasizes the key role played by specification in the design process. Several types of specifications are needed at each design level, relating to various parts of the artefact A, and also to the tests required by the quality assurance process. The figure shows the LEVEL INPUT specification as the initiator of the nth subphase of the design process, and the design resulting from the level n subprocess. The level n design contains the following types of information:

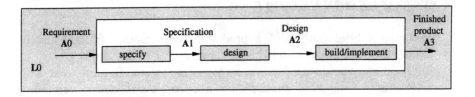

Figure 4.14a: Outline of the product/process model of the supply phase of
the life history of a system

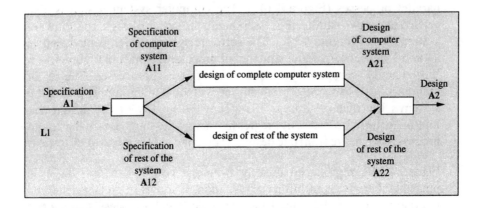

Figure 4.14b: Top level model of the system design process

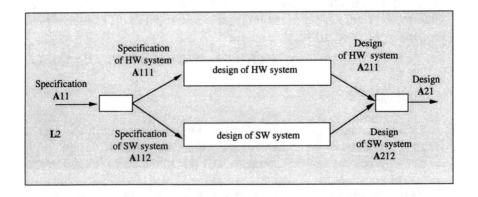

Figure 4.14c: Second level model of the computer system design process

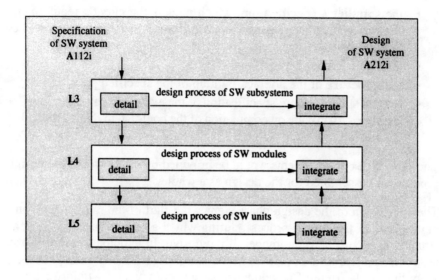

Figure 4.14d: Further levels of the model of the software design process

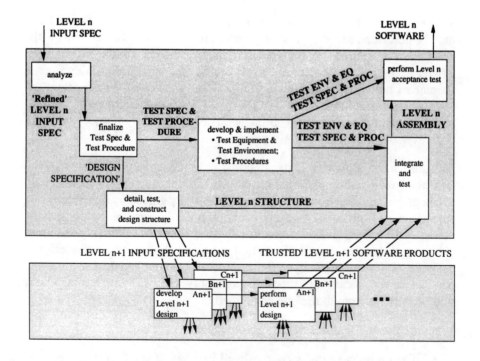

Figure 4.14e: The product/process model of a level within the design process

- the **supplier's specification**, which models the client's requirement, but also incorporates any additional attribute which the supplier wishes to impose;
- the set of **part-specifications**;
- the **structure** of the artefact, built up of the specified parts;
- the **test specification**: specification of the evidence which the supplier is obliged to provide on completion of the contract, to demonstrate that the artefact complies with the client's specification.

The same types of documents would be provided by each level of the design process, and assembled for the design as a whole.

The conclusion of the design phase signals the start of the implementation / build phase of the supply process, during which a real-life artefact is constructed in accord with the model of the design. In anticipation of this, the design must incorporate evidence of the practical realizability of the design, according to the state of development of technology, and within the given resource requirements and timing constraints.

4.4.2 Design strategies

The regime which would bring about the conditions of Figure 4.14a to 4.14e may be described as a design strategy of 'part-by-part validation'. This is the strategy of a cautious realist who sets more stringent conditions than are strictly necessary. In practice, a correct design may have been achieved even if some of the components failed to meet their prescribed specifications, so long as others compensated for this, and the artefact met the over-all requirement of the relation $S_P{\leftarrow}S_D$. The models of the design process given in Figures 4.14 amount to such a strategy.

An alternative approach is the optimist's design strategy, which may be described as 'it-will-be-all-right-on-the-day'. Here design is not regulated by a structure of rigorous quality tests, but is informally guided by the prescriptive specification. It proceeds by using common sense in partitioning the specification of the whole, choosing structures and components intuitively. Informal checking and review is usually carried out along the way; nevertheless, the designer is forced to wait until the process is complete to see how the descriptive specification of the whole assembly will work out, hoping that it will be close enough to the one prescribed.

What are the key features of the two strategies?

- The optimist's strategy serves well in cases of small projects, but has several major disadvantages, which rule it out for projects which are complex and sizeable, and where the penalty for regression is high.

* The prescriptive specification of the whole artefact, as the designer's only reliable guide in choice of the the atomic components and structures, can become too remote for design decisions at the level of detail, and yet it is at such detailed levels that the atomic components are selected.

* The component behaviours are tightly coupled throughout the design, and any change in a part, even at the extremities of the structure, may set off a ripple effect, reaching the behaviour of the whole artefact. This means that design problems and requirement variations cannot easily be localized: even a small component change, design modification, or adjustment of client's demands, can have a major effect on the design of the system as a whole.

* Throughout the design process, the designer, and the management, are left in suspense as to the wisdom (or otherwise) of structural and component choices. As a consequence, managerial decision-making relies less on fact and design skill than on rhetoric and skillful advocacy, and some good projects may be prematurely curtailed, while doomed projects may carry on, throwing good money after bad.

• The realist's part-by-part strategy has great advantages:

* This strategy gives a degree of autonomy for the design of each component. It sets individual prescriptive specification targets for each part, deliberately distributing among the parts the volume of the specification space.

* Within the bounds of their prescriptive specification space, the descriptive specification of components may vary, without the change affecting the system specification. Hence the strategy allows the designer to limit inter-component coupling.

* The strategy gives the designer managerial control over the design process. It allows the placing of independent subcontracts, while also providing assurance (through formal proof or validation) that the design at the higher levels is correct. It gives to designer and management peace of mind that the whole process will succeed, as long as each subcontract is fulfilled.

* The strategy enhances the robustness of the design. If clients' requirements change, the nominal properties of each component may be altered, within the limits of the design space. If the changes are more extensive, the design structure affords the identification of

those components whose specifications have been stretched beyond their limits, preserving the integrity of the rest of the design.

- Although for large and safety-sensitive projects the realist's is the only feasible strategy, one must not lose sight of some major risks:

  * As design progresses, and as contingency factors are built in at every stage of the top-down design process, a 'squeeze' is imposed on the specification domain. This may escalate project costs, due to the over-specification of atomic components.

  * The realist's strategy may turn out to be a pessimist's strategy, leading to the unwarranted failure of the whole project, whereas a more liberal design regime, which allows for tighter coupling and inter-component compensation, may have succeeded.

4.5 Specification and design of processes

Until now, attention has been focused on cases where the referent was an artefact, and specification and design yielded *products* as versions of this same referent. The requisite characteristics of the referents were specified by instantaneous properties. Whenever processes entered into the discussion, they were considered as the means of obtaining the next version of the referent, rather than being referents in their own right.

In many cases the requirement is not satisfied by a product whose attributes of interest are immediately perceivable; instead, it calls for a process, whose task is to deliver a response to a stimulus. The product/process model of Figure 4.1a provides the guide to the specification and design of entities of this kind, and we reproduce the model here in Figure 4.15, for convenience.

4.5.1 Life history of processes

Figure 4.16 is a product/process model which represents the *life history process* of a referent entity v, which is a process. The entities V_0 to V_5 are intermediate products of the complete life history process, relating to the referent v.

- The 'Requirement' V_0 of v is a product of the client's mind.

- The 'Specification' V_1 of v models the desired process in the form given in Figure 4.15, and also stipulates the desired property variable values.

- The 'Design' V_2 of v defines the structure of v as a process, composed of subprocesses, such that V_1 should be satisfied.

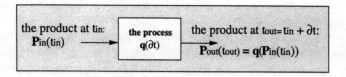

Figure 4.15: The product/process model

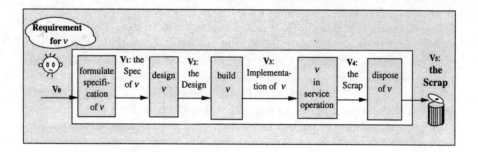

Figure 4.16: The life history of a process

- The 'Implementation' V_3 of v is an artefact which carries out the required service operation of v.

- The process v 'dies' when the artefact which implements v is finally disposed of.

4.5.2 Specification of processes

How can we specify an entity such as the process **q** of Figure 4.15?

Section 4.3.1 defines the specification of a product as a black-box model:

- the black-box model of the product $M_B = (\textbf{Prop}, \textbf{R}_{Prop})$, and
- the set of variable values for the properties of the set **Prop**.

It would be prudent to make full use of the definition of products when devising a way of specifying processes. However, the product/process model of Figure 4.15 tells us that a process **q** is a fundamentally different entity from a product: it is the means of transforming some product $\textbf{P}_{in}(t_{in})$ into another product $\textbf{P}_{out}(t_{out})$ over the set time $\delta t = t_{out} - t_{in}$. Processes are harder to specify than products, because, while the product can (and should!) be specified context-independently, without reference to the process which brought it about, a process must be specified with reference to its products, in the context of the triple: $(\textbf{P}_{in}(t_{in}), \textbf{P}_{out}(t_{out}), \textbf{q})$.

We demonstrate two ways of using product specifications as the means of specifying processes, referring to these, respectively, as the 'snapshot' approach, and the 'causal' or 'transfer function' approach. Within the latter, we discriminate between processes with and without memory. The examples are kept very simple, and the mechanisms are well known, but here we show how the product/process model makes explicit some of the important implied process characteristics.

The 'snapshot' approach to process specification

Consider some referent entity Π, such as a manufacturing company, an oil pipeline in service, or an AND gate carrying out logical operations. The essence of such referents is the process which they perform, and a process model would display their most revealing representation. Nevertheless, each of these referents can also be viewed as products. The product model of Π takes a snapshot of the input and output products entering and exiting these referents at the time when the snapshot is taken. Since the process needs finite time to respond to an incoming product, the output at a given time instant may have no visible causal link with the input which appears at that same time. Hence the action of the process may not be deducible from a snapshot, but the snapshot can have its own uses, as we shall see presently.

The snapshot of Π can be specified as any other product: a black-box model, together with the corresponding property values. The black-box model of Π will take the usual form:

$$M_B (\Pi) = (\mathbf{Prop}(\Pi), \mathbf{R}_{Prop}(\Pi)),$$

where $\mathbf{Prop}(\Pi) = (\mathbf{Prop}_{in}, \mathbf{Prop}_{out})$ are the pair of property sets describing the products $\mathbf{P}_{in}$ and $\mathbf{P}_{out}$, respectively,

and $\mathbf{R}_{Prop}(\Pi) = (\mathbf{R}_{Prop\ in}, \mathbf{R}_{Prop\ out})$, the pair of relation sets for $\mathbf{P}_{in}$ and $\mathbf{P}_{out}$, with common time signature.

We shall give no further attention to such models here, other than noting that, if judiciously applied, they can have good practical uses: they may find their way into process specifications, and a sequence of such snapshots might be a useful indicator of trends. A snapshot model of the rate of flows into and out of the pipeline would show whether pressure in the line is building up or reducing. The snapshot of the input and output voltage levels of the AND gate may make no logical sense (the inputs may have changed, while the output still corresponds to the previous value), but the snapshot model would indicate instants of high power dissipation during the transient. Care must be taken, however, not to be confused by the snapshot model, and not to use it beyond its sphere of applicability. A couple of examples are mentioned to illustrate this danger.

- Some programming languages admit expressions in the form:

$$X = X + 1 \dots\dots\dots\dots\dots\dots\dots\dots\dots\dots\dots\dots\dots\text{Eq 4.2}$$

If the syntax were to indicate the time delay separating the X on the left of the '=' sign from the X on the right, the expression would be a good user-friendly model of the assignment process, communicating the programmer's intention to the reader of the program. The syntax of PASCAL achieves this by using the 'becomes equal' symbol ':=' to designate the *process*. In a product/process model such as Figure 4.2a, the responsibility is entrusted to the *product* symbols rather than to the process: the distinct time signature of the two instances of 'BeatenEggs' allows for differences in the properties of products of the same name. The syntax of equation 4.2 uses neither of these means, and the result reads as nonsense.

- The classical theories of hazard-free design of logic circuits were worked out decades ago, when propagation delays were negligible compared with the processing times of the gate circuits of the day, and the delay in the signal transmission process could be ignored. Now that logic circuits are implemented in microelectronic technology, the assumption is no longer valid, but the theory is still taught and applied in many places.

Transfer functions in process specifications

Consider once again the entity Π, which is now modelled as the process of Figure 4.15. The process $\mathbf{q}$, which maps the product $\mathbf{P}_{in}(t_{in})$ into the product $\mathbf{P}_{out}(t_{out})$, is referred to as a 'transfer function', and is given in equation 4.3.

$$\mathbf{P}_{out}(t_{out}) = \mathbf{q}\ (\mathbf{P}_{in}(t_{in})), \text{ where } t_{out} = t_{in} + \delta t \dots\dots\dots\dots\dots\dots\text{Eq 4.3}$$

Equation 4.3 describes the product $\mathbf{P}_{out}(t_{out})$ as a function, such that its specification is deducible if $\mathbf{q}$ and the specification of $\mathbf{P}_{in}(t_{in})$ is known. Conversely, one may formulate the specification of the process $\mathbf{q}$ in terms of the specification of its two products.

$$\text{Spec}(\mathbf{q}) = (\ \text{Spec}(\mathbf{P}_{in}(t_{in}), \text{Spec}(\ \mathbf{P}_{out}(t_{out})) \dots\dots\dots\dots\dots\dots\text{Eq 4.4}$$

such that products $\mathbf{P}_{in}(t_{in})$ and $\mathbf{P}_{out}(t_{out})$ are specified in the known manner.

Recall now that a product is a unique entity, specified by its black-box model which includes its time signature, and the set of its property values. Very few processes are intended to work just once, on a single unique product, and equation 4.3 restricts itself to these. Most processes are required to act several times, as called upon during their service period, and operate on items selected from a whole genus of products. This means that $P_{in}(t_{in})$ in equation 4.3 must be generalized to the product species Q_{in}: the input domain of q, comprising the set of *all* products within q's time span and property domain.

We use the classification scheme devised in Section 4.3, denoting the genus Q_{in} composed as the set of species of input products:

$$Q_{in} = \{ Q_{in1}, Q_{in2}, \ldots \},$$

where   a set of input products $Q_{in\,j} \in Q_{in}$ will comprise all input products $P_{in\,j}(t_{in})$ in q's domain which differ only in their time signature, but otherwise have identical properties.

Let the common property set of products in $Q_{in\,j}$ be denoted by In_j. Then an individual input product $P_{in\,j}(t_{in}) \in Q_{in\,j}$ will be identified by its (time-less) property set In_j, and its (individual) time signature t_{in}. The input product domain Q_{in} of q is now definable as:

- the set of all input property sets $In = \{ In_1, In_2, \ldots \}$,
- the time domain of input products $T_1 \le t \le T_2$, where T_1 and T_2 are delimiters of the time domain

with a corresponding output range Q_{out} of:

- the set of output property sets $Out = \{ Out_1, Out_2, \ldots \}$,
- the time span of output products $(T_1 + \delta t) \le t \le (T_2 + \delta t)$.

Now we may use the generic representation of input products to re-write equation 4.3 to describe the general case, when q may be called upon to operate repeatedly, and act on different sorts of products:

$$Q_{out}(t + \delta t) = q\,(Q_{in}(t)), \ldots\ldots\ldots\ldots\ldots\ldots\ldots\ldots\ldots\ldots\text{Eq 4.5}$$

Here t is constrained within bounds $T_1 \le t \le T_2$, the time domain of q's service operation,

    δt is the process delay caused by q; in general, the time delay δt cannot be assumed constant, but will vary over the whole of q's domain.

    $Q_{in}(t) = \{ Q_{in\,1}(t), Q_{in\,2}(t), \ldots \}$ is the product genus comprising the input product domain of q,

    $Q_{out}(t) = \{ Q_{out\,1}(t), Q_{out\,2}(t), \ldots \}$ is the product genus of the range of output products generated by q,

    all products $P_{in}(t_{in}) \in Q_{in\,j}(t)$ have identical properties, differing only in their time signature,

    and, likewise, all products $P_{out}(t_{out}) \in Q_{out\,i}(t)$ have identical properties, differing only in their time signature.

In equation 4.5, **q** is a 'potential process' which can be instantiated at any time in the time domain of $\mathbf{Q_{in}}$. The service life of the process **q** would be composed of 'actual' processes: versions of **q**, created for various instances of products in $\mathbf{Q_{in}}(t)$. The service life of **q** will be describable as a product/process model whose atomic input processes have inputs $\mathbf{P_{in}}(t_{in})$ in **q**'s domain $\mathbf{Q_{in}}(t)$, and generate outputs $\mathbf{P_{out}}(t_{out})$ in **q**'s range $\mathbf{Q_{out}}(t)$.

THE 'AND GATE' EXAMPLE

This example demonstrates the principles of the transfer function approach to process specification.

The referent is the two-input AND gate of Figure 4.17a which performs the process **g**. It is part of some equipment with a stipulated service life of five years, say. The time delay of the gate is constant, c.

The genus of input products of the process has three dimensions, defined by the triplet of variables $(\mathbf{x_1}, \mathbf{x_2}, t\,)$,

where $\mathbf{x_1}, \mathbf{x_2}$ are single-digit Boolean scalar variables defining the input property set **In**,

and   t is a discrete variable denoting elapsed service time with respect to the start of the service operation of **g**, defining the time signature of a particular product $\mathbf{P_{in}}$ with properties $\mathbf{In_j}$ within **g**'s domain such that $0 \le t \le 5$ years.

The output products of $\mathbf{Q_{out}}$ are then defined in closed form as:

$$\mathbf{Q_{out}}\,(t+c) = \mathbf{g}(\mathbf{Q_{in}}(t)) = AND(\mathbf{x_1}, \mathbf{x_2}, (t+c)) \quad\dots\dots\dots\dots\text{Eq 4.6}$$

where   c is a constant with the dimension of time, giving the time delay of the gate, for example, $0.02\mu sec$.

An actual process may come about by applying to **g** the input product defined, for example, by the triple (T, F, $(t_0+100\ \mu sec)$), yielding the definition of the output as the pair (F, $(t_0+100.02\mu sec)$, with t_0 as the start of the gate's service life, measured by the real-time clock.

In Figure 4.17a, the 'actual' process is shown, acting on the product $\mathbf{P_{in}}(t_{in})$. Figure 4.17b gives the 'potential' (generic) process, with time kept as variable. In both figures the usual AND gate symbol replaces the rectangular process box. Figure 4.17c is the traditional truth table of the AND gate: this is a time-independent representation of the causal relationship of the corresponding input/output product pairs.

END OF AND GATE EXAMPLE

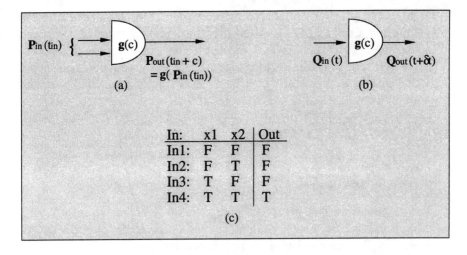

Figure 4.17:
'Actual' and potential product/process model and specification of an AND gate

State transition processes

In most cases the transfer function of some process **q** cannot be defined uniquely on the basis of the input product alone, unless the previous history of the process is also known. Such processes, which have memory, are called 'state transition processes'.

Figure 4.18a shows the product/process model of a state transition process **q** which operates on $P_{in}(t_{in})$ and produces $P_{out}(t_{out})$. **q** is composed as a structure of two parallel processes **q'** and **q"**. One of these, **q"**, generates the external process output $X_{out}(t_{out})$. The other, **q"**, creates the auxiliary 'state output' $Y_{out}(t_{out})$ whose role is to keep a record of past history. Both **q'** and **q"** have the same composite input, comprising the external input $X_{in}(t_{in})$ and the 'state' input $Y_{in}(t_{in})$, generated by the previous instant of process operation. Since **q'** and **q"** are parallel processes, their input and output products in the product / process model are synchronized.

For discrete systems with finite memory, the usual representation of a state transition process is by the Mealy or Moore model. These show the causal link between system states, outputs and transitions [8]. The extension of these models to real-time is a subject of continuing interest (see e.g. [9]). We now demonstrate here briefly the product/process model approach.

Traditional models of state transition processes put the auxiliary process in a feedback loop around the main process. In the product/process model feedback loops are disallowed, thus making it explicit that time cannot move

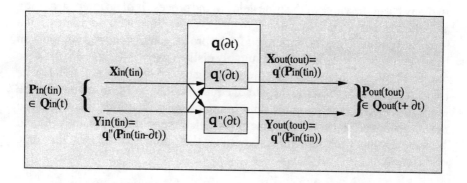

Figure 4.18a: The operation of a state transition process

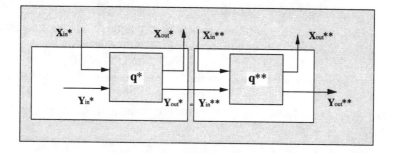

Figure 4.18b: Chaining of two stages of the state transition process

backwards; hence instances of the state transition process are concatenated, chaining state inputs along the sequence of time frames of **q**'s repeated operations, to form the complete state transition process. Figure 4.18b shows this for two time frames, denoted by '*' and '**', respectively.

The input/output relation of equation 4.3 applies also to state transition processes, with the interpretation of **q** given in Figure 4.18a. Equations 4.7a and 4.7b define each of the parallel processes which make up the state transition process, generating the process output $\mathbf{X}_{out}$ and the state output $\mathbf{Y}_{out}$, with appropriate time signatures.

$$\mathbf{X}_{out}(t_{out}) = \mathbf{q}' \; (\mathbf{X}_{in}(t_{in}), \mathbf{Y}_{in}(t_{in})), \quad\text{.....................................Eq 4.7a}$$

$$\mathbf{Y}_{out}(t_{out}) = \mathbf{q}'' \; (\mathbf{X}_{in}(t_{in}), \mathbf{Y}_{in}(t_{in})), \quad\text{....................................Eq 4.7b}$$

where $t_{out} = t_{in} + \delta t$,

and, in addition, $\mathbf{Y}_{in}(t_{in}) = \mathbf{q}'' \; (\mathbf{P}_{in}(t_{in} - \delta t), \mathbf{Y}_{in}(t_{in} - \delta t))$.

In many cases the value of the time delay of the process is of no interest, as long as a clear and explicit distinction is made between adjacent time frames at the input and output. Here the arguments of the process output and state output can be suppressed for simplicity, but a suitably distinct name is assigned to the state output, emphasizing its essential timekeeping role. A possible notation for the transfer function is shown in the pair of equations 4.8a and 4.8b, where Y is the input provided by the previous time frame of the state output, and Y^+ is the newly generated state output, sometimes called the 'next state output'. Y and Y^+ have distinct time signatures; otherwise their properties may be the same or different.

$$\mathbf{P}_{out} = \mathbf{q}\ (\mathbf{X}_{in}, \mathbf{Y}),\ \dots\dots\dots\dots\dots\dots\dots\dots\dots\dots\text{Eq 4.8a}$$

$$\mathbf{Y}^+\ = \mathbf{q}''\ (\mathbf{X}_{in}, \mathbf{Y}))\ \dots\dots\dots\dots\dots\dots\dots\dots\dots\text{Eq 4.8b}$$

The generic form of equation 4.5 also applies to state transition processes. In the equation

$$\mathbf{Q}_{out}(t+\delta t) = \mathbf{q}(\mathbf{Q}_{in}(t)),$$

$\mathbf{q}$ is now interpreted as the parallel process pair $\mathbf{q}'$ and $\mathbf{q}''$. Abstracting away from time, we have two sets of properties of input products:

- $\mathbf{In} = \{\mathbf{In}_1, \mathbf{In}_2, \dots\}$ referring to the property values of various sets of products in $\mathbf{X}_{in}$, the same as for processes without memory, and
- $\mathbf{D} = \{\mathbf{D}_1, \mathbf{D}_2, \dots\}$ designating the property values of sets of products of $\mathbf{Y}_{in}$.

There are two sets of time-unrelated output properties:

- $\mathbf{Out} = \{\mathbf{Out}_1, \mathbf{Out}_2, \dots\}$ as before, and
- $\mathbf{D}^+ = \{\mathbf{D}_1^+, \mathbf{D}_2^+, \dots\}$ designating the sets of output products of $\mathbf{Y}_{out}$.

Where only the property values are of significance, it is often useful to model the product sets by their property values, allowing $(\mathbf{In}_i, \mathbf{D}_j)$ and $(\mathbf{Out}_k, \mathbf{D}^+_l)$ to stand for instances of $(\mathbf{X}_{in}, \mathbf{Y})$, $(\mathbf{X}_{out}, \mathbf{Y}^+)$, respectively. With this in mind, we may now define the finite state transition process $\mathbf{q}$ as follows:

Definition

Let $\mathbf{q}$ be a state transition process, and let $\mathbf{D}_i$ and $\mathbf{In}_j$ be species of state and input products of $\mathbf{q}$, respectively, such that members of $\mathbf{D}_i$ and members of $\mathbf{In}_j$ differ only in their time signature, but otherwise have the same properties. If the complete state product set $\mathbf{D} = \{\mathbf{D}_1, \mathbf{D}_2, \dots\}$, the input product set $\mathbf{In} = \{\mathbf{In}_1, \mathbf{In}_2, \dots\}$ and the output product set $\mathbf{Out} = \{\mathbf{Out}_1, \mathbf{Out}_2, \dots\}$ are all finite then $\mathbf{q}$ is called a *finite state process* FSP.

End of definition

For simple processes without retentive memory, the state equation 4.7b is not required, and $\mathbf{q}''$ does not exist. This means that the process of Figure 4.18 reduces to that of Figure 4.15, and equation 4.7a and 4.7b degenerate to equation 4.3. The simplified version of the method for specifying state transition processes also serves for specifying such processes.

Forms of process specification

The specification of the process $\mathbf{q}$ might be given in at least the following ways:

- Defining $\mathbf{q}$ formally, in *closed form* as a function, or as relations expressed *declaratively*, as a system of rules of deduction over a given domain of input products, such that the specification of any possible output product $\mathbf{P}_{out}(t_{out})$ in $\mathbf{Q}_{out}(t)$ could be derived by applying $\mathbf{q}$ to the specification of the corresponding input product $\mathbf{P}_{in}(t_{in})$ in $\mathbf{Q}_{in}(t_{in})$.

- Defining the function of $\mathbf{q}$ by full or partial *enumeration,* giving the set of all (**In, Out**) pairs for some or all of the valid domain of $\mathbf{q}$.

- Defining $\mathbf{q}$ *procedurally*, as a controlled sequence of processes, such that any valid instance of $\mathbf{Q}_{in}$ should yield a valid instance of $\mathbf{Q}_{out}$.

- Giving some *informal* or *semi-formal* textual description of $\mathbf{q}$, perhaps complemented by drawings, graphs, etc.

- Defining $\mathbf{q}$ *operationally,* by constructing a simulator or prototype artefact Ω to serve as a dynamic model of $\mathbf{q}$, and generate any model instance of $\mathbf{Q}_{out}$ in response to any valid instance of $\mathbf{Q}_{in}$. Here the specification is implicit in the workings of the artefact Ω.

All five process specification methods find use in practice, and we refer to some of them in later chapters. The arguments in favour of formal and explicit specifications apply to processes as much as they apply to products, and we confine the use of the term 'specification' to such representations of requirements for potential processes.

As noted before, the literature contains many ways of specifying state transition processes. Here we use the model of Figure 4.18a, and the life history process of Figure 4.16, to construct our definition.

Definition

Let $\mathbf{V}_0$ be the requirement for some process ν, to be satisfied by a finite state process $\mathbf{q}(\delta t)$. Then the *specification* $\mathbf{V}_1$ of $\mathbf{q}(\delta t)$ is the pair of parallel processes $\mathbf{q}'(\delta t)$ and $\mathbf{q}''(\delta t)$ of common input domain $\mathbf{Q}_{in}(t)$.

The primary process of $q(\delta t)$ is $q'(\delta t)$, which generates the external output $X_{out}(t+\delta t)$. The auxiliary process of $q(\delta t)$ is $q''(\delta t)$. It generates the state output $Y_{out}(t_{in}+\delta t)$, which supplements the external input product for the next instantiation of $q(\delta t)$.

End of definition

An input product to q, $P_{in}(t_{in}) \in Q_{in}(t)$, comprises:

- the external input product $X_{in}(t_{in})$ supplied to q and characterized by the set of properties $In_j \in In$;

- the state input product $Y_{in}(t_{in})$, supplied by q itself, characterized by the set of properties $D_i \in D$;

- the time signature $t = t_{in}$, common to input products X_{in} and Y_{in}, where, in accord with the definition of the finite state process, In and D are finite.

If the processes $q'(\delta t)$ and $q''(\delta t)$ of $q(\delta t)$ are specified by enumeration then both will take the form of a set of triples: the external input and state input pair from sets In and D, and the product output generated by each process:

- pairs $(In_j \in In, D_i \in D)$ which $q'(\delta t)$ maps into an external output product with properties $Out_k \in Out$

- pairs $(In_j \in In, D_i \in D)$ which $q''(\delta t)$ maps into a state output product with properties $D_i^+ \in D^+$.

Enumeration is *complete* if it extends to the full complement of pairs: all members of the cross-product of the sets In and D.

Frequently $q(\delta t)$ is specified in a table of four columns: the input column pair for elements of In and D, and the output column pair for corresponding elements of Out and D^+.

Processes in their environment

The need for creating a process arises from the environment, and the specification must represent the operation of the process in context. Figure 2.4 in Chapter 2 modelled the interaction of the referent system and its environment in a closed world. We now indicate briefly how that world view fits in with the product/process model developed in this chapter.

Figure 4.19a repeats the 'closed world' model of Figure 2.4 for the case where the 'referent system' is the process q, and the 'environment system' is the process e. The two processes are contained in a 'closed world'. They interact with each other throughout the life history of their universe by exchanging generic products, and have no contact with any other process.

To create the product/process model of **q** and **e** as an 'actual' process, we must recall, yet again, the key features of products, processes, and product/process models: (i) products are instantaneous, (ii) an 'actual' process operates on a product once and once only, (iii) processes always involve a passage of time (and, conversely, passage of time implies the existence of a process), and (iv) processes are linked to each other by products.

Since **q** and **e** operate together, they are parallel processes, with synchronized time frames. Figure 4.19b shows the actual process for the case when neither process has retentive memory. **q** and **e** are instantiated repeatedly, and since the universe is closed around the two processes, neither process has any other role but to serve the other. The process sequence of Figure 4.19b is initiated by the start-off event of process **q**'s service operation phase. The product pairs which enter a process at subsequent instantiations will have progressive time signatures (time marches on). In general, the products entering a given process will have different properties, and hence the duration δt of the processes is variable; however, the two processes operate in parallel, and remain synchronized throughout their life.

The model is more elaborate when **q** is a state transition process. Here **q** is itself composed of two parallel processes **q'** and **q"**, as in Figure 4.18a. In addition to its obligation to serve **e**, at each instantiation **q** must provide an output for maintaining its own integrity. In general, **e** will also be such a state transition process, and the product/process model will be of the form as in Figure 4.19c. The notation follows that of Figure 4.18a: at each subsequent instantiation **q** and **e** pull in the state output generated by the previous instantiation. The exceptions are $Y_{11}(t_0)$ and $Y_{21}(t_0)$ which have no predecessors. In practice, their values are supplied at the start, by the 'initialization' of the processes.

4.5.3 Design of processes

As for products, the design of processes involves choice, adds detail, and creates structure. If the resulting structure V_2 is a product / process model which performs some transformation, and if V_2 meets the specification V_1 for a process required for a process v, then V_2 is a suitable design for v.

We met a specific case of process design when converting a potential process into the service life of an actual process. Selecting the property variable values and time signature of input products created a set of specific instances of the potential process, which were actual atomic processes. Assembly of these atoms into a product/process model creates an actual process. If this meets the specification V_1 for some process, then the structure V_2 is the design of that process.

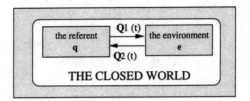

Figure 4.19a:
The referent process **q** in interaction with the environment process **e**

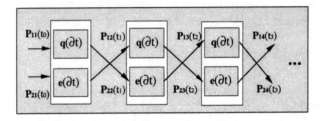

Figure 4.19b:
The product/process model of a 'memory-less' referent process **q**
in interaction with the memory-less environment process **e**

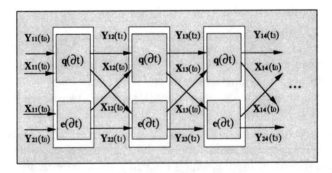

Figure 4.19c: The product/process model of the state transition referent
process **q** in interaction with the environment process **e**

In general, the design of a process will amount to a product/process model, decomposable into atomic process components, forming a multi-level nested structure of series/parallel composition.

Definition

Given a finite state process specification V_1 for process ν, then the design of the process, V_2, is a product/process model over specified atomic finite state processes.

The structural model takes the standard form $M_S(V_1) = (\textbf{Comp}, \textbf{R}_{Comp})$, where the set **Comp** comprises the specifications of the atomic components in V_2, and the relation set $\textbf{R}_{Comp}$ is the structure of the process.

End of definition

4.6 Summary

The *requirement* for an **artefact** is the collection of its desired attributes. The *specification* of an artefact is an instance of its black-box model, stipulating a value for each member of the property set, such that the artefact should meet the requirement. The *design* of an artefact is a structural model, resolved to implementable atomic components of known black-box behaviour, which meets the specification. Requirement, specification and design are all *versions* in the life history of the artefact. The life history can be represented by a product/process model: an acyclic directed graph, in which nodes and arcs stand for (instantaneous) products and processes of finite duration. The artefact of interest may be abstract or tangible, living or inanimate. It may be a bespoke entity, or member of a family of mass-produced items. Specification and design are models of the artefact, but they can also be viewed as products in their own right. For any but the rarest and simplest artefacts, specification and design are composite, multilevel processes. In course of the design process, the model of the specification is reified, by added detail, into a structure. Part-specifications and structural specifications are tested throughout, to guard the design process against error, non-compliance, and loss of efficiency caused by undetected errors.

In many cases the requirement relates to a **process**, rather than to an artefact. A process is defined as a *transfer function*: a mapping of the input product of the process into the output product. To assure that the mapping is indeed a function (i.e. that a unique output product corresponds to each input product), the definition of *state transition* processes (processes with memory) must include a state function which keeps track of the past history of the process. The *specification* of a process defines the process function. Most processes operate repeatedly, and can be called upon to act on a variety of products; hence their specification must define their time and input product domain, as well as the mapping which they perform. The *design* of the process is a product/process model which imposes a structure over the specifications of atomic processes, such that the resulting process conforms with the specification.

The concepts of artefact, product, process, specification and design are defined in this chapter, explained through their models, and are illustrated by several simple examples. This provides foundations for devising model-based measurement schemes of specification and design.

4.7 References

1 Shorter Oxford Dictionary. 1987.

2 Boehm B W (1981): "Software engineering economics". Prentice Hall.

3 IEE (1989): "Guidelines for assuring testability". Publication catalogue.

4 W M Turski, T S E Maibaum (1987): "The specification of computer programs". Addison Wesley.

5 Halsbury's Laws of England (4th Ed.) (1973-1987) Vol 4, Page 579. Butterworths

6 Jevons W S, and Nagel E (1958): "The principles of science". Dover Publications.

7 Kaposi Associates (1992): "Design for Testability". Course Document GR/5/92.

8 Hopcroft J, Ulman J (1979): "Introduction to automata theory". Addison-Wesley.

9 Avnur A (1990): "Finite state machines for real-time software engineering". Computing and Control Engineering Journal, November, Vol 1, No. 6, pp 273-278. IEE.

5   MEASURES OF SPECIFICATION AND DESIGN

5.1 Introduction

Chapter 3's introduction to model-based measurement amounted to a warning: no number or symbol can characterize an entity meaningfully, unless there is a formal model which establishes the context. Recalling Lord Kelvin's motto, the same message can be put more explicitly: "What is not measurable, make measurable *through its formal model*". Once a formal model exists which represents selected attributes of a referent, the model displays these as measurable properties. In addition, the model will often reveal further properties which point to unidentified inherent attributes, enriching the picture of the referent and enhancing understanding.

Chapter 4 gives us the definition of specification models for representing specifications and designs of artefacts and processes. Here we show how these models lead naturally to their measurement-based characterizations. In devising measures, we follow the principles and methods outlined in Chapter 3, mapping entities of the 'WORLD OF MODELS' into corresponding entities in the 'WORLD OF MEASURES'.

5.2 Measures and measurement schemes for product specifications

5.2.1 Object-oriented measures of product specifications

In Chapter 4, the specification of a product $\mathbf{P}(t)$ was defined by the property values of its black-box model. This means that the product specification is made up of two parts:

(1) the product's black-box model, in the form:

$$\mathbf{M_B(P}(t)) = (\mathbf{Prop, R_{Prop}}) \qquad \text{...Eq 5.1}$$

where $\mathbf{M_B}$ is the black-box model of the product $\mathbf{P}(t)$,

$\mathbf{Prop} = \{\mathbf{prop}_1, \mathbf{prop}_2, ...\mathbf{prop}_n\}$ is the set of properties of $\mathbf{P}(t)$, and R_{Prop} is the relation set over the properties, including the mandatory individuality relation rp_1 which carries the time signature,

(2) a set of $(n+1)$ data items: the **n** property measures and the timing measure of the product.

The form of the product specification will be the same whatever the artefact. Hence one may build a generic measurement scheme, as in Figure 3.6, for the object-oriented characterization of all product specifications.

The measures of the scheme arise directly from the definition of product specification, which in turn relies on the definition of the black-box model. Accordingly, the measurement scheme will include two sorts of measures:

- *property measures*, denoted by **m**: the set of measures for each element **prop**$_i$ in the property set **Prop**;

- *the time signature*, denoted by **t**: a measure of real time, carried by the individuality relation rp_1 in the relation set R_{Prop}.

The black-box model carries further information, which must also be captured in measures. We know from Chapter 2 that an irredundant black-box model is defined in terms of independent, orthogonal properties. For such black-box models, the relation set has a single member, the individuality relation rp_1, which is already measured by the time signature. When the black-box model is redundant then R_{Prop} will contain further relations, and to characterize them, the measurement scheme has an additional class of measures:

- *relation measures*, denoted by **y**: tree measures of the relations

     $rp_i \in R_{Prop}$

where each relation is represented as a structure of homogenes.

In practice, it has been found useful to define two further measures, indirect measures which concisely represent formal attributes of the black-box model, rather than detailed information about the product. These are:

- two *set measures:* general measures defined over all sets, applied here to the sets which comprise the black-box model: **Prop** and R_{Prop}. These are denoted by **p** and **r**, respectively.

The object-oriented measures of a product specification assemble into a five-tuple in the form of equation 5.2:

$$M_{obj}(P(t)) = (\, p, r, m, y, t \,)............................Eq \ 5.2$$

These are now discussed in turn.

Set measures

The 'size' of a set is measured by its cardinality: the number of elements which it contains. The notations $|Y|$ or $C(Y)$ are used to denote the cardinality of some arbitrary set Y. Adopting the former notation, the measurement scheme of an artefact specification will contain the counting measures:

$$p = |\mathbf{Prop}|, \text{ and } r = |\mathbf{R}_{Prop}|.$$

For an orthogonal property set, $r = |\mathbf{R}_{Prop}| = 1$; in other cases the value of the measure indicates the extent of redundancy of the specification.

The size $p = |\mathbf{Prop}|$ of the property set is a useful indication of the 'complexity' of the specification of the artefact, but the measure must be used with care. As we have seen, the specifier has wide options over the definition of property variables, and these options extend even to the number of property variables for representing a given attribute. For example, in a crude avionics system, the attribute 'direction' is defined as the single discrete variable ß, with a domain of 36 points, $0 \leq ß \leq 36$, measuring angle with relation to North; in another system, engineered by the same company, the same attribute is defined differently, not by one but by three variables: two single-digit Boolean variables designating N/S and E/W respectively, and a discrete variable µ with a domain of 9 points, representing the subdivision of each quadrant to the same accuracy.

Property measures

$m = \{ m_1, m_2, ... \}$ is the complete set of the measures for the set of properties $\mathbf{Prop} = \{ prop_1, prop_2, ...\}$ in the black-box model of product $\mathbf{P}(t)$.

If the product is a physical entity with well-understood attributes then its property variables will be defined within domain-specific theories, and direct or indirect measures for them will be provided within the unit system of classical metrology. In general, such measures are not unique, and the specifier has many options. As an example, Chapter 3 showed that the attribute 'width', fully understood and measurable, could be represented by a continuous or a discrete length variable, and, for either of these, many different scales of measurement could be chosen. The notion of 'size' is intuitively understood, but there is a variety of property variables which can represent it. The choice is even wider when the referent arises from a developing application domain, where there are no established concepts, and the specifier has definitional obligations and options. Nevertheless, once the property model is defined, metrology can be entrusted with the task of delivering the property measures. As discussed in Chapter 3, each property variable is defined by *type* and *domain.* In the specification each property is given a measure on an appropriate scale, and, additionally, each quantitative measure carries *dimension* as well as *magnitude.*

Relation measures

Relations of $\mathbf{R}_{Prop}$, with the exception of the individuality relation $\mathbf{rp}_1$, are usually given in the form of mathematical or logical expressions which define a property variable of **Prop** in terms of others. The expressions can be modelled as tree structures of homogenes, and they can be characterized by tree measures. It follows from the notion of structural models discussed in Chapter 2 that a single-level tree can also be drawn up for the individuality relation $\mathbf{rp}_1$, with the whole entity as its root, the atomic components as its leaves. In this case, $\mathbf{y} = \{\ \mathbf{y}_1,\ \mathbf{y}_2,\ ...\ \}$ is the complete set of tree measures for characterizing the relation set $\mathbf{R}_{prop} = \{\ \mathbf{rp}_1,\ \mathbf{rp}_2,\ ...\ \}$.

Time signature

$\mathbf{rp}_1$ designates the instant of the product's existence. The time signature $\mathbf{t}$ is a measure of real time.

5.2.2 Utility measures of product specifications

Utility measures of an entity are defined in Chapter 3 as explicit functions over the set of its object-oriented property measures. The utility attributes of product $\mathbf{P}(t)$ must conform to the criteria of *measurability* listed in Chapter 3, and the set of utility measures $\mathbf{M}_{ut}(\mathbf{P}(t))$ of product $\mathbf{P}(t)$ are a class of its specification measures. Such measures will be in the form:

$$\mathbf{M}_{ut}(\mathbf{P}(t)) = \{\mathbf{u}_1,\ \mathbf{u}_2,\ ...\ \mathbf{u}_i,\ ...\} \quad ....\text{Eq } 5.3$$

where $\mathbf{u}_i$ stands for the ith utility measure of $\mathbf{P}(t)$, defined as
$\mathbf{f}_i\ (\ \mathbf{p},\ \mathbf{r},\ \mathbf{m},\ \mathbf{y},\ \mathbf{t}\ )$,

$\mathbf{f}_i$ is a function over the set of $\mathbf{P}(t)$'s object-oriented measures, designed to reflect the user's notions of the ith utility attribute,

and   symbols $(\ \mathbf{p},\mathbf{r},\mathbf{m},\mathbf{y},\mathbf{t}\ )$ have the meaning explained previously.

There are at least two types of utility measures, depending on their role: *merit measures* which articulate the specifier's value system, for example, to order members of a class of artefacts on a scale of merit, and *aggregate measures* which assign some established, widely accepted index to a product class. Technically, both are utility measures of the form of equation 5.3, but while merit measures are entirely subjective and local to their designer, the aggregate measures have broader constituency, and some claim to a consensus.

A possible example of an aggregate measure could be the 'complexity' of specifications. Complexity is an elusive attribute, and it is a sign of the wide interest in it that many attempts have been made in the past to define it. One of the intended uses of the 'complexity' of the specification for some artefact is in predicting design and development costs, but, as yet, no convincing model has been put forward for connecting predictor to predicted. It is reasonable to assume however that an artefact whose specification prescribes many properties, and whose property domain is wide, will be harder to describe, and more expensive to test comprehensively, than one with a small domain. Even if these 'complexity' features do not capture some inherent natural attributes of the artefact itself, they will reflect on the ease with which the specification is comprehended and documented. It also reflects on the client who is ordering the artefact: a fussy client will prescribe many properties in minute detail, and will be harder to satisfy, than the customer who is content with a simpler, less detailed specification. One may have some sympathy with the definition of a 'complexity measure' which aggregates the information content of the **m** measures of the product specification.

5.2.3 Measurement scheme for product specifications

The measurement scheme for products will then comprise the property variable definitions and relations of the black-box model of the product, together with the specification's object-oriented measures, $M_{obj}(P(t))$, and utility measures $M_{ut}(P(t))$. For convenience, we combine these into the 'specification measure':

$$M_{spec}(P(t)) = (M_{obj}(P(t)), M_{ut}(P(t))).$$

The measurement scheme will be built on the foundation of direct and indirect measures, as shown in Section 3.2, Chapter 3.

5.2.4 Measures of generic product specifications

In Chapter 4, we introduced generalizations of products (see Figure 4.13). The specification of an *individual product* $P(t)$ carries the full complement of property values and the time signature to qualify its black-box model. A *product species* Q has a subset of these data items, whereas a *product genus* G is void of all data, even of the time signature of its black-box model. The difference in the detail of the specification may be shown by defining object-oriented measures for product species and product genus, as well as individual products:

$$M_{obj}(P(t)) = (p, r, m, y, t) \dots\dots\dots\dots\dots\dots\dots\dots\dots Eq\ 5.4a,$$

with meaning of symbols as defined in Section 5.2.1,

$$M_{obj}(Q) = (\mathbf{p, r, m', y, t'}) \dots\dots\dots\dots\dots\dots\dots\dots\dots\dots\dots\text{Eq 5.4b}$$
where $\mathbf{m'}$ is a subset of the property measure set $\mathbf{m}$,
and $\mathbf{t'}$ may be empty, or else it is the time signature $\mathbf{t}$;

$$M_{obj}(G) = (\mathbf{p, r, \emptyset, y, \emptyset}) \dots\dots\dots\dots\dots\dots\dots\dots\dots\dots\text{Eq 5.4c,}$$
where $\emptyset$ signifies the empty set.

These equations highlight the attribute of the 'resolution' of a specification, and its converse: the 'size' of the population of products to which the specification applies. We denote the property of resolution by $\mathbf{z}$.

Definition

Let S be a product species whose black-box model has $\mathbf{n}$ properties ($\mathbf{p=n}$), and let the specification of the species include $\mathbf{z}$ number of value definitions. Then the resolution, $\mathbf{z}$, will be a positive scalar integer defined over the domain:
$$0 \le z \le (p+1),$$
with $\mathbf{z} = 0$ corresponding to the whole genus $\mathbf{G}$,

$\mathbf{z} = (\mathbf{p+1})$ corresponding to an individual product $P(t) \in G$,

and $0 < \mathbf{z} < (\mathbf{p+1})$ corresponding to species $\mathbf{Q}$ of $\mathbf{G}$.

End of definition

It may sometimes be useful to measure resolution by a real scalar $\mathbf{s}$, obtained by normalizing $\mathbf{z}$ to $(\mathbf{p+1})$. Now $\mathbf{s}$ will have the domain:

$$0 \le s = z/(p+1) \le 1,$$
with $\mathbf{s} = 0$ corresponding to the whole genus $\mathbf{G}$,

$\mathbf{s} = 1$ corresponding to an individual product $P(t) \in G$,

and $0 < \mathbf{s} < 1$ corresponding to the various species $Q \in G$.

The definition treats 'product genus' and 'product' as extreme cases of 'product species', distinguishing them by the value of the resolution measure. In its integer form $\mathbf{z}$, the resolution measure enumerates the two subclasses which make up the $(\mathbf{p+1})$ property variables of the black-box model: the subclass with $\mathbf{z}$ members for which measures have been defined, and the remaining subclass of $\mathbf{p+1-z}$ which are left as variables. The resolution measure ($\mathbf{z}$ or $\mathbf{s}$) imposes a partial order over the complete set of species which comprise $\mathbf{G}$, and quantifies the generality (or specificity) of a given specification. Resolution is an attribute which has been found to have importance when choosing among various specification options.

Using now the notion of 'product species' in the widest sense, extending it to the extreme cases of product genus and individual product, we may compact equation 5.4a, b and c, and include **z** into the single six-tuple:

$$M_{obj}(Q) = (p, r, z, m, y, t) \dots\dots\dots\dots\dots\dots\dots\dots Eq\ 5.5,$$

where   the symbols have the meaning just defined, except that **m** now stands for the set of *actual* property measures which **z** calls for, and **t** may be empty.

5.2.5 Measurement scheme for generic product specifications

The six-tuple object-oriented measure of equation 5.5 forms the 'data base' on which to build the set of utility measures of the product species:

$$M_{ut}(Q) = \{u_1, u_2, \dots u_i, \dots\} \ . \dots\dots\dots\dots\dots\dots\dots Eq\ 5.6$$

where u_i stands for the ith utility measure of the product species **Q**, defined as f_i (**p, r, z, m, y, t**)

and f_i is a function over the set of **Q**'s object-oriented measures, especially designed to reflect the user's notions of the ith utility attribute, and symbols (**p, r, m, z, y, t**) have the usual meaning.

The measurement scheme for product species will then comprise the property definitions and relations of the black-box model of the species, together with the specification measures:

$$M_{spec}(Q) = \{ M_{obj}(Q), M_{ut}(Q), \} \dots\dots\dots\dots\dots\dots Eq\ 5.7.$$

where $M_{obj}(Q)$ and $M_{ut}(Q)$, are defined in equations 5.5 and 5.6,

and   equation 5.7 incorporates the whole product genus and individual products as special cases of product species, in accord with equation 5.4a, b, c.

5.2.6 Measures of product design

In Chapter 2, equation 2.6 gave the general form of a structural model. Later, Chapter 4 showed design as the process of 'instantiation' of specifications by component selection and imposition of structure. Where the entity of interest is a product, the design takes the form:

$$M_S(P(t)) = (Comp, R_{Comp}),$$

with   **Comp** as the set of components of **P**, defined by their black-box model and common time signature,

and R_{Comp} as the structure which the design imposes on the component black-box model set,

the behaviour of the design satisfying the specification of the product **(P(t))**.

The same applies to the design of a product species or a whole product genus. The design of any of these types of systems is a multilevel hierarchical structure, ultimately resolved into atomic components of known behaviour, the design being valid if the black-box behaviour manifested by the structure meets the system specification, as seen in the discussion of structural modelling in Chapter 2.

The measures characterizing the system design are built up from the specification measures of the system components, together with the measures of the system structure. If the system is a multilevel hierarchy, it will contain both atomic and non-atomic components. Each non-atomic component, and the system itself, will be a substructure over lower-level components, the whole design being a nesting of substructures. The directed graph measures of the structure R_{comp} are used as indicators of the probability of error in the design, and as predictors of system properties, such as maintainability, on the observation that complicated structures are error-prone and hard to repair.

For verification of the correctness of the design against the prescribed specification, one must obtain the behaviour of the system described by the design. One possibility is to implement the design hoping that it is correct, and validate the system properties directly, by measuring the finished product and comparing it with the prescribed specification. The realist designer would wish to rely less on hope and more on evidence, and would seek to gain confidence in the correctness of the design before releasing it to the implementation phase of the life history.

In principle, the measures of the system's black-box behaviour can be derived from their design in two ways, as shown in Figure 5.1. From the black box viewpoint the structure itself is immaterial, but it is essential as the means of deriving the system's black-box properties. Both routes of Figure 5.1 start from the same design: a multilevel structure over known atomic components. The steps taken are as follows:

Route 1:
- measure the design by measuring each of the atomic components of the system, choosing a symbol system which preserves the composition operators of the design structure;
- compose the measures of the system specification from measures of its design.

Route 2:
- compose the system specification from the system design;
- measure the system specification.

The measures must be the same, regardless of the route taken to obtain them. In practice, usually some intermediate arrangement is adopted, as will be shown later. The detail of measurement strategy is left to the measurer's convenience, provided that the invariance of measures is maintained.

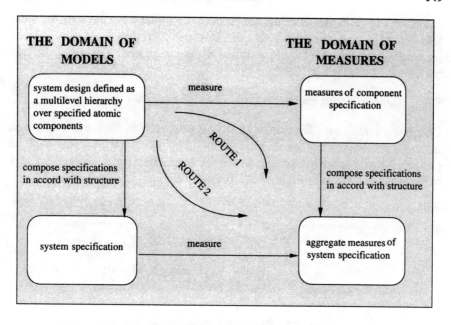

Figure 5.1: Composition of design measures for characterizing
a system as a structure of its atoms

5.3 Process specification measures

We noted in Chapter 4 that processes are more elaborate entities than products: they must be defined in terms of their products. Sophisticated methods for modelling and measuring the functions, resource requirements and capability of processes [1] have been developed in branches of control engineering, manufacturing and process engineering. Process models and measures are currently devised for software, but these relate to the development process, rather than to the action performed by the software in operation.

Construction of a general measurement scheme for processes would require extensive research, and the exploration of the subject is beyond the scope of this book. Instead, we proceed by using simple examples of some specific process classes which have interest in themselves, and, through these, demonstrate the principles of constructing model-based measurement schemes.

Finite state processes have been the subject of attention in Chapter 4. A scheme for their model-based measurement has been worked out in earlier research, and the scheme has been demonstrated on some examples, under sponsorship by BT (see e.g. [2]). Although that case study concerns a simple telecommunication system, it is too extensive to be presented here, but a summary account has been published in [3]. In this chapter we pursue, in detail, a 'toy example' of a finite state process, and use it as a vehicle for explaining the main features of the measurement scheme. A case history of a comprehensive measurement scheme of quite a different kind comprises Chapter 6 of this book, and is detailed in Part 2.

5.3.1 Measures and measurement scheme of 'actual' processes

Consider first the specification of an atomic process q, to be used just once in its entire service life for transforming a specified product $P_{in}(t_{in})$ into another product $P_{out}(t_{out})$. Such a process has been seen many times before, but it is reproduced yet again in Figure 5.2, for convenience.

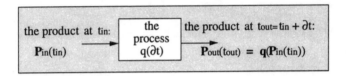

Figure 5.2: The 'actual' single-shot process q

Process q is a 'single-shot' process: it has neither past nor future, and needs no memory. Since q is atomic, its product/process model has a single process component. Its specification is a *structure,* composed of two elements: the specification of the pair of products $P_{in}(t_{in})$, $P_{out}(t_{out})$. The structural relation of the product/process model, rq, say, indicates that the two products belong to the same entity, namely, the process q. Measures of the single-shot atomic process will be compiled as follows: the object-oriented measure is a triple:

$$M_{obj}(q) = (\ M_{spec}(P_{in}(t_{in})),\ M_{spec}(P_{out}(t_{out})),\ m_{rq}\ ),$$

where m_{rq} is a structural measure, representing the process as a single-level tree with two leaves.

All atomic processes will have the same single-level binary tree structure, and hence m_{rq} may be omitted without loss of information. Then the specification measure of q becomes the pair:

$$M_{obj}(q) = (\ M_{spec}(P_{in}(t_{in})),\ M_{spec}(P_{out}(t_{out}))\ ).$$

Utility measures $M_{ut}(q)$ can now be devised in the usual way over the object-oriented measure $M_{obj}(q)$. The specification measures $M_{spec}(q)$, is then:

$$M_{spec}(q) = (M_{obj}(q),\ M_{ut}(q)).$$

Assume now that there are no utility measures defined for q, nor for either of the products of q. With this simplification, the specification measure of q reduces to a pair of six-tuples:

$$M_{spec}(q) = (\ M_{obj}(P_{in}),\ M_{obj}(P_{out}))\ \dots\dots\dots\dots\dots\dots\dots\dots\text{Eq 5.8}$$
$$= ((\ p_{in},\ r_{in},\ z_{in},\ m_{in},\ y_{in},\ t_{in}),\ (\ p_{out},\ r_{out},\ z_{out},\ m_{out},\ y_{out},\ t_{out})),$$

where the meaning of the symbols follows from previous explanations.

Since for actual processes the time signature and all property variables are defined by value, $z_{in} = p_{in}+1$ and $z_{out} = p_{out}+1$. Thus, in equation 5.8 both z measures can be omitted without loss, and equation 5.8 can be written as a pair of five-tuples. These constituents allow us to construct the simplified measurement scheme of the single-shot atomic process.

Where some of the simplifying assumptions do not hold, the measurement scheme for the process will be more elaborate, but the model of Figure 5.2 still applies, and the measurement scheme will be constructed as above.

Where the actual process is not atomic but is composed as a structure of atomic processes, we must return to the expression of a structural model of a system in equation 2.6, and interpret it for the particular case. The structural model will now be in the form:

$$M_S(q) = (\textbf{Comp},\ \textbf{R}_{Comp}),$$

with   **Comp** as the set of component processes of q,

and $\textbf{R}_{Comp}$ is the structure which the design imposes on the component process set.

In principle, one would construct process measures by following the path described for products in Section 5.2.6. In practice a general measurement scheme for processes is yet to be devised, but models of processes are products of the modelling process, and hence they are amenable to measurement.

5.3.2 Measures of potential processes

Potential processes are processes which can operate on product *species* rather than individual products. Typically, a process is designed to operate repeatedly, members of the species being stretched out in the time domain. The process may need to act on a range of products of different properties. The process may have a variety of functions, the choice of function changing with the properties of the incoming product, and/or controlled by a signal which forms part of the incoming 'product'. A process may serve several functions simultaneously, and may be an elaborate structure which can 'pipeline' in accord with some protocol, for example, starting to work on a newly arriving product while completing work on a previous arrival. The process will seldom be atomic: it will usually be composed as a structure of subprocesses. It is the role of the modeller to find a suitable representation of the process so as to understand and predict its properties and results, and for all such processes the 'actual' life history process will be described by a product/process model of appropriate series/parallel structure, rather than by the single-shot process of Figure 5.2.

We discuss here the simplest of potential processes: the atomic potential process $q(\delta t)$ of Figure 5.3a, used repeatedly on members of the product species Q_1, where any product $P_{1i}(t_{1i}) \in Q_1$ differs from any other product $P_{1j}(t_{1j}) \in Q_1$ only by its time signature, but otherwise has the same properties. In practice, $q(\delta t)$ may be some manufacturing process function acting on the work piece of a production line.

One strategy for devising a measurement scheme for a potential process may be to resolve it to an actual process. Figure 5.3b shows the actual process model of the service operation of the potential process of Figure 5.3a. The process acts on composite products: in addition to the incoming and outgoing work pieces, there is a stream $\{P_{21}, P_{22}, ... \}$ of 'products' linking the atoms of the actual process, each product $P_{2i}(t_i)$ signalling that the process is ready to accept the next work piece $P_{1i}(t_i)$. The process measures for q will be composed by first devising measures for one of the members of the sequence of Figure 5.3b, and then composing measures with the aid of the single-level tree which represents the structure of the actual process. If the potential process q of Figure 5.3a is not atomic then each of its repetitions in Figure 5.3b resolves into further detail, and the tree is a multilevel structure of repeated substructures.

Another strategy, often more convenient in practice, is not to convert the potential process model to an actual process model, but to develop a measurement scheme from the potential process model directly, taking account of the variation in the timing and properties of the product species. We discuss some simple instances of this for state transition processes.

5.3.3 Measures of state transition processes

Take the case where q is a finite state process (FSP) with **In** and **D** modelling the complete external input and state input product species, and **Out** and D^+ representing the corresponding species of external and state output products. As we noted in Chapter 4, the specification for the FSP can be compiled in a four-column 'state transition table', enumerating the process specification fully or partially, giving instances of (**In**, **D**) input pairs and corresponding values of (**Out**, D^+) output pairs. Section 5.4 will show many such examples.

Another way to present the same information about a state transition process is in a 'state transition diagram': a directed graph which abstracts away from real time, and shows the paths of possible 'journeys' which may be taken during the life of any actual process. There are various forms of such diagrams.

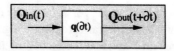

Figure 5.3a: The potential process **q**, acting on the product species $Q_{in}(t)$ to produce the product species $Q_{out}(t+\partial t)$

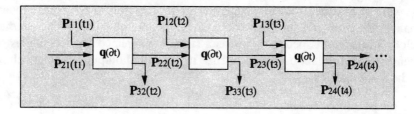

Figure 5.3b: The actual service operation process implemented by $q(\partial t)$

The one we use later in the chapter assigns a node to each state of the state set **D**, and a *directed arc* to each step which may be taken between states in response to an input product in the set **In**. If the cardinality of the state set **D** is $|D| = k$, then the state transition diagram has **k** number of nodes. A complete k-node directed graph would have k^2 number of arcs; the graph which models a given FSP will have a subset of the arcs of a complete graph, and hence its transition process set **K** will have up to k^2 members. Measures of state transition processes may then be defined through properties of their state transition table and diagram models, as we shall show presently.

Because of its generality and conciseness, the state transition diagram is a popular model for representing finite state systems. Practical systems have too many states for the state transition diagram to be directly applicable, but methods have been devised to extend the scope of these models and their methods by abstracting groups of states into clusters, and representing each super-state cluster in a node of a directed graph called the 'state chart' [4].

5.4 FSP models and measures in practice

5.4.1 Introduction to specification measures of finite state processes

We introduce the specification measures of finite state processes through the extended 'light bulb example'. Features are added gradually, in five stages (version A to E), and the simplicity and familiarity of the process helps the reader to concentrate on the ideas underlying a measurement scheme.

The operation of the light bulb in all versions of the example is described by a state transition diagram. This diagram is a directed graph: the type of product/ process model described in Figure 4.1b. Nodes of the graph correspond to products (or product species), and arcs to processes. In a state transition diagram of our example, there are two sorts of atomic processes: (i) *transient* processes, which move the light bulb from one state to another (for example, switching it ON if it was OFF), and (ii) *state maintenance* processes, which keep the light bulb in a given state (for example, ON if it was already ON). Looking at the state transition diagram in another way, a transient process changes the properties of the product, such that the input and output product belong to different species, whereas a state maintenance process makes no change in the product properties, hence its input and output products belong to the same species.

The measurement scheme for a finite state processes is based on building a *classification scheme* [5] of its products and processes. The product/process model defines the entire population of states and state transitions of the FSP, and leads to criteria for their classification. The direct measures of the measurement scheme are scalar integers which enumerate the population of states, state transitions and their sub-classes.

The classification system comprises three parts, and these correspond to the three kinds of direct measures of the measurement scheme: Part 1 classifies the states of the FSP, and Part 2 and Part 3 classify state transitions by their syntactic and semantic characteristics, respectively.

The models, classification schemes and measures apply equally to all versions of the example, but we pay particular attention to versions C and E. For these, we derive state and state transition classes and measures, and show them in tabular form. The two versions are sufficiently similar to demonstrate the generality of the classification and measurement scheme, and sufficiently different to allow comparison between instances of systems of this kind. The reader is invited to complete such tables for the other versions of the example.

In addition to developing the concepts and procedures of a model-based measurement scheme for finite state processes, the examples indicate how such a system may be used to assure quality.

LIGHT BULB EXAMPLE – VERSION A

Figure 5.4 is the state transition diagram which describes the service operation phase of a family of light bulbs in ordinary domestic use. In this simplest version of the example we assume that the operation of light bulbs is adequately described by just two states, and hence the state set $\mathbf{D}$ has two members: $\mathbf{D}_{ON}$ and $\mathbf{D}_{OFF}$. The light bulb is a physical entity, and each of its states refers to a 'product' with a distinct set of properties measurable by the system of classical metrology (such as the quantity of light being emitted, the amount of electrical power being dissipated, the surface temperature of the glass envelope, etc.).

The light bulb belongs to a large family of mass-produced artefacts whose black-box model at the start of their service operation is barely distinguishable, and hence their state transition model will be the same; however, each will have its unique history of use. During its service history, an individual light bulb may visit each of the two states many times. If, for any reason, it should be necessary to represent the *actual* life history of the light bulb in real time, then the state transition model must be expanded into a time sequence of distinct items of the light bulb as an individual artefact.

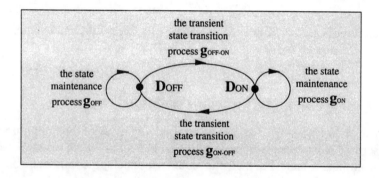

Figure 5.4: State transition diagram representing the service function of a light bulb as a two-state process

END OF VERSION A

LIGHT BULB EXAMPLE – VERSION B

The model of Figure 5.4 is not unique: it captures just one part of the life history, and is just one of many possible state transition models which may be constructed to describe this part of the life history. Figure 5.5 represents a FSP which models wider aspects of the life history process: the normal service life as well as the failure of the light bulb. The FSP has now three states: $D=\{D_{ON}, D_{OFF}, D_{FAIL}\}$. Once the light bulb fails, there is no possibility of repair, and hence no return from the FAIL state. The FAIL state is maintained indefinitely, until the user declares the end of the light bulb's life history. The state transition model of Figure 5.5 represents more than one member of an artefact family: it maps out two routes which any

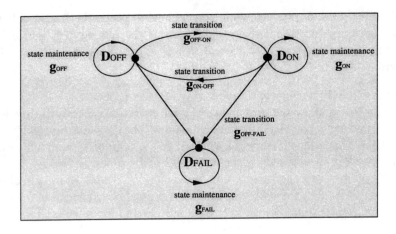

Figure 5.5: The light bulb in service, modelled as a three-state process

individual bulb will follow to the failure state. Each individual of the family will take one or another of the routes to $\mathbf{D_{FAIL}}$, but not both.

END OF VERSION B

5.4.2 The finite state process in the closed world

Figures 5.4 and 5.5 show how the operation and failure of the light bulb may be described by state transitions, but give no indication of the *cause* of movement from state to state. For that, one requires to specify the finite state process in interaction with its environment.

Consider now the 'closed world' system described in general terms in Figure 2.4. In the present case, shown in Figure 5.6, we assume that both the 'target object' **T** and the 'environment' **E** are finite state processes, so that the whole 'closed-world system' **C** also amounts to a finite state process, made up of **T** and **E** as an interacting pair. In the example of Figure 5.5, the 'target object' **T** would be the light bulb, while the 'environment' **E** would incorporate the user who controls the state of the light bulb by means of a switch. **C** will be defined as $\mathbf{C} = (\mathbf{D_C}, \mathbf{G_C})$, where $\mathbf{D_C}$ is the set of states of **C**, and $\mathbf{G_C}$ is the set of *state transitions* defined over $\mathbf{D_C}$.

STATE CAPACITY

If the target object **T** has **n** states (the state set $\mathbf{D_T}$), and the environment **E** has **m** states (the state set $\mathbf{D_E}$), then the closed world system **C** is a FSP which has *state capacity* for **n x m** states. The state capacity is defined by the number of items in the cross-product $\mathbf{D_T} \times \mathbf{D_E}$. The state set of a given model $\mathbf{D_C}$ may comprise any subset of this.

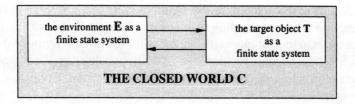

Figure 5.6:
The two-part closed world as a pair of interacting finite state systems

STATE TRANSITION CAPACITY

Assume now that the cardinality of D_C is k. The complete directed graph has a set of k^2 arcs, defined by the cross-product of the nodes $D_C \times D_C$, and, in principle, this is the *theoretical state transition capacity* of the FSP. However, working with the full set of state transitions would create excessive complexity in practice. Hence most specifiers will adopt a modelling method to achieve contraction: some discipline which imposes simplifying assumptions, so as to rule out certain *classes* of state transitions. For a given state set, the chosen method will leave only a subset of directed arcs G_{Cmax}, corresponding to the 'admissible transitions'. The cardinality of the set G_{Cmax}, denoted by $|G_{Cmax}|$, is called the *actual state transition capacity* of the k-state FSP. The state transition diagram of a given FSP representing the closed world C will map out the actual transition set G_C which utilizes a subset of the state transition capacity; thus, the following relation holds between the size of the actual transition set, the actual transition capacity and the size of the state set:

$$|G_C| \leq |G_{Cmax}| \leq k^2.$$

We return to the light bulb example to illustrate.

LIGHT BULB EXAMPLE- VERSION C

Consider again the simple operation in Figure 5.4. Viewed in the closed world paradigm, the state transition diagram is that of Figure 5.7. The environment E amounts to a human-operated switch. In the example, each of the two components has two states, ON and OFF. The model does not describe the possible failure of the light bulb; regardless of whether the switch and bulb are ON or OFF, the states of the FSP are always 'operational'.

Figure 5.7 shows the states and state transitions of such a closed world, formed by the light bulb with its switch. States are numbered, and each state carries the label pair 'switch state, lamp-state'. There are two state maintenance processes, and in the notation of the figure the self-loops of these have been subsumed into the corresponding nodes (states 1 and 3). States with self-loops are often referred to as *stable states;* those without are called *transient states.* Figure 5.8 is the state table for version C of the light bulb example, corresponding to the state diagram of Figure 5.7. We refer to this as Part 1 of the description of the operation.

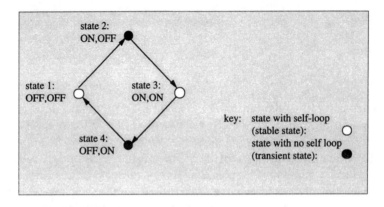

Figure 5.7: The working light bulb in a closed world

	$D_T = \{OFF, ON\}$ $D_E = \{OFF, ON\}$	COMMENTS
ordinal number	$D_E \times D_T$	
1	OFF, OFF	stable, operational
2	ON, OFF	transient, operational
3	ON, ON	stable, operational
4	OFF, ON	transient, operational

Figure 5.8: Part 1, the state table of the working light bulb and switch

This FSP has a state capacity of 2 x 2 = 4 states. All of the capacity is used by the state set D_C.

There could be 4 x 4 = 16 processes linking the state set D_C, including all possible self-loops. Only 6 of these are needed to describe the operation of the light bulb, and hence only these are included in the path set G_C of the working system.

The FSP of Figure 5.7 uses 37.5% of the 4-state system's state transition capacity.

END OF VERSION C

In general, unused capacity signifies inefficiency or error. Could the system have been described by a simpler model, with fewer states? Or is a system of this capacity needed, but perhaps the modeller has forgotten some of the requirements? Or perhaps the unused capacity is only an illusion, and the effectively available capacity is fully utilized?

USING STATE CAPACITY FOR ENHANCING SAFETY

In many situations some of the capacity is reserved for enhancing the reliability and safety of the operation of the FSP. For example, some state transitions are sensitive to accuracy of timing, and if this accuracy cannot be guaranteed, the system may malfunction. The safe thing to do, if possible, is to avoid such hazardous state transitions by imposing operational or design rules. These rules reduce the system's effective capacity, but increase safety and reliability.

In our example we adopt a system of operational rules which build in two assumptions. They are given below, together with their justification.

(i) Since no two events can be precisely simultaneous, only one of the two closed-world components is allowed to change state at a time.
(ii) Once a state transition process chain is initiated, it must be allowed to reach conclusion, and the system to settle in a maintained state before a new state transition event may be started.

Collectively these two have been referred to as the *'fundamental mode assumptions'* of classical asynchronous finite state systems. A similar code guides conversation in a civilized society, where no two people speak at the same time, and a speaker is allowed to conclude a sentence without interruption. (Note that the same communication protocol is imposed on the boys by themselves in William Golding's "Lord of the Flies".) Many real-life situations call for more sophisticated operational protocols, but the fundamental mode assumptions suffice for our present purposes, and we keep to them for the rest of this series of examples.

Interpreting these assumptions for VERSION C of the lightbulb example, note that they:

(i) rule out direct links between nodes in the diagonals of the rectangle of Figure 5.7, and
(ii) force a follow-on step from state 2 to state 3, and also from state 4 to state 1, prohibiting the reverse path to the originating node.

RECOGNIZING IMPLICIT LIMITATIONS OF THE USABLE STATE CAPACITY

The real-world referent may itself imply assumptions which exclude some state transitions. In our example, the assumption is that in the relationship between target and environment, the switch is the master and the lamp is a slave. The slave has no initiative, always responds to command, and can only change state if bid to do so. This rules out arcs from state 1 to state 4, and from state 3 to state 2, leaving only the transitions in Figure 5.7.

LIGHT BULB EXAMPLE – VERSION D

Figure 5.9 is a new model of the closed-world system of lamp and switch. The lamp can now exist in three states: ON, OFF and FAIL. The controlling switch has two states; ON and OFF, as before.

The state set D_C' of this new closed-world system C' has capacity for 3x2=6 states. Only five of these are utilized in the state transition model. Four of the five used states are 'operational' states, corresponding to the working system. The fifth state is a 'non-operational' error state, involving the failed bulb. The model makes no use of the other possible error state OFF, FAIL.

In state transitions between the four operational states, the lamp remains a slave of the switch, but it now has the initiative to transfer to the FAIL state.

END OF PART D

In proposing the model of Figure 5.9, the specifier has made at least two implicit assumptions:

(i) Once the light bulb has failed, the state of the switch cannot change.

(ii) The light bulb can only fail from the ON state.

Let us check what these assumptions mean for our example, and see what general conclusions may be drawn.

STATE COMPLETENESS

Compare versions C and D, using the Figures 5.7 and 5.9. Version C has 2 environment states and 2 system states, giving a state capacity of 2x2=4. The state transition diagram shows that all states are used. Version D uses 5 states out of the total of 6: an 83% state utilization. A model such as this, with less than 100% state-utilization, is given the name *state-incomplete*.

It is generally unsafe to leave any state out of the model. It is advisable to impose quality checks which detect state-incomplete models, and demand their revision. In the case of the light bulb example there is nothing

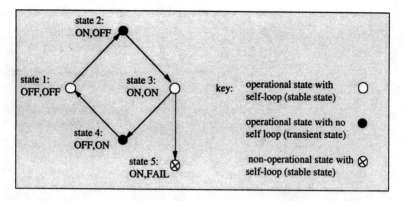

Figure 5.9:
The closed-world FSP of a light bulb liable to failure, and its controlling switch

to prevent the user of the light from operating the switch, even after the bulb has failed; hence the error state OFF,FAIL can occur in practice.

LIGHT BULB EXAMPLE – VERSION E

Figure 5.10a shows a possible state-complete model of our FSP, with Figure 5.10b as the corresponding state table. The model shows that the two non-operational states (5 and 6) have a mutual link. Of course, once the bulb has failed, the switch has no effect on the light, but the operation of the switch transfers the system from one failed state to the other. Of the six states, four are stable (with self-loops), and two are transient.

                                              END OF VERSION E

TRANSITION COMPLETENESS

A complete 4-node directed graph has 4x4=16 arcs. Hence a 4-state system, such as VERSION C in Figure 5.7, modelled by a 4-node directed graph, has a theoretical state transition capacity of 4x4=16. Not all of these are possible: some are ruled out by our assumptions. As we have seen, VERSION C uses all state transitions allowed by our modelling method. We call such systems *'transition complete'*.

Compare VERSION E in Figure 5.10a, modelled by a six-node directed graph.This has a theoretical state transition capacity of 6x6=36. The chosen modelling method rules out 2 state transitions for each stable state, and 5 transitions for each transient state. This leaves an actual state transition capacity of 18. The model uses 11 of these, including the self-loops of the stable states: a 61% transition utilization. Models, with less than 100% state transition utilization, are called *transition-incomplete*.

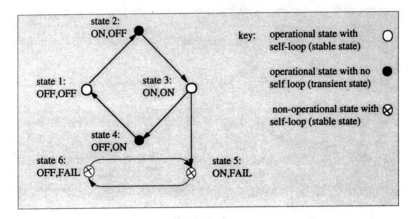

Figure 5.10a:
A state-complete model of and switch. The bulb can only fail when switched ON.

	$D_E = \{OFF, ON\}$ $D_T = \{OFF, ON\}$	COMMENTS
ordinal number	$D_T \times D_E$	
1	OFF, OFF	stable, operational
2	ON, OFF	transient, operational
3	ON, ON	stable, operational
4	OFF, ON	transient, operational
5	ON, FAIL	stable, non-operational
6	OFF, FAIL	stable, non-operational

Figure 5.10b: Part 1: The state table of failure-prone light bulb and switch

As we noted before, it is advisable to impose quality checks on FSP models to detect transition-incompletness, and revise such models to account for each state transition explicitly, by stating the modelling assumption which eliminates them. Reverting to VERSION C,

- one exit from each of the 2 stable states is ruled out by the lamp's lack of initiative to move out of stable states;

- one exit from each non-operational state is ruled out by lack of repair, which prevents return to the operational domain;

- one exit is missing from state 1 because of assumption (ii), that the bulb can only fail from the ON state.

Using the notions of state-complete and transition-complete systems, one may now devise a procedure for generating 'safe' specifications, and impose a procedure for assuring the quality of specifications. Among the important aspects are that all states and state transitions are explicitly accounted for, and redundant states which form no part of the user requirement are employed in the interest of enhancing safety and reliability.

5.4.3 Classification of states and state transitions

Consider again the closed-world system C of Figure 5.4, defined as $C=(D_C, G_C)$. Assume that C is state-complete and transition-complete. We offer criteria for classifying states and state transitions of such a system, as preliminary to characterization by measurement. State classification is presented in full; transition classification only in outline. A more comprehensive treatment is given in [3].

Classification of states

We have already identified two types of states:

- States with self-loops, called *stable states*, form the set D_{Cs}. These states occur repeatedly, sustained by a state maintenance process, unless and until a state transition process overrides the state maintenance.
- States without self-loops, called *transient states*, form the the set $D_{Ct} = D_C - D_{Cs}$.

We have also met another criterion of classification:

- *Operational states* are part of the normally working system, forming the set $D_{Cop} \subseteq D_C$.
- *Non-operational* states are all those states which take no part in the correct functioning of the system. The set of non-operational states is $D_{Cnop} = D_C - D_{Cop}$.

Applying these two criteria to states gives us four distinct categories:

1. *Operational stable states*, defined by the set $(D_{Cop} \cap D_{Cs})$.

2. *Operational transient states*: part of the working system, encountered on route between operational stable states. These are defined by the set $(D_{Cop} \cap (D_C - D_{Cs}))$.

3. *Non-operational stable states*, defined by the set $(D_C - D_{Cop}) \cap D_{Cs}$. These are *irrecoverable error states* (states 5 and 6 in Figure 5.9b), or stable states corresponding to the process of repair.

4. *Non-operational transient states* are in the set
$$((\mathbf{D}_C - \mathbf{D}_{Cop}) \cap (\mathbf{D}_C - \mathbf{D}_{Cs})).$$
They either lie on route to or from non-operational states, or are *redundant* states of the state-incomplete system, not called for by the specification. There is a quality requirement on the specifier to take account of all such undefined states, but judicious use of these states offer possible enhancement of the system, either for increased safety or increased functionality. When used in such a manner, non-operational transient states become reclassified as operational transient states.

These state classifications are marked in the 'COMMENTS' column of Figures 5.8 and 5.10b.

Classification of state transitions – 1

We have already observed that two types of *state transition processes* occur in a FSP, distinguishable from the syntax of the model:

- *State maintenance* processes (also called *steady state* or *stable state* processes) sustain a given stable state, and shown as self-loops in the state transition model.
- *Transient* state-transition processes signify action, and modelled by arcs which link distinct states of the state transition model.

  Depending on their *cause*, transient processes divide into two sub-classes:

  * *'forced'* or *'reactive'* transients, and
  * *'free-will'* or *'spontaneous'* transients.

  A forced transient state transition (Part 2 of the operational description) is a response of one of the constituents of the two-part closed world system to the stimulus of the changed state of the other part.

  A free-will transient state transition (Part 3) is, at least to some degree, unpredictable. Typical cases occur in systems which incorporate human beings, or are used in modelling random mechanisms, such as failure in hardware systems.

We choose a design which imposes an *order of precedence* on the three classes of processes just identified (state maintenance, free-will state transient, and forced transient). The order is the following:

(1) *Forced* transient state transitions have the highest priority.
(2) *Free-will* transient state transitions have intermediate priority
(3) *State maintenance* transitions have the lowest priority.

It follows that free-will transient processes are only possible from a stable state.

The examples of Figure 5.9 and 5.10a show all three types of state transition processes, and lead to general observations.

- All exits from stable states, except self-loops, are free will.
- All exits from transient states are forced.

In the light bulb example of Figure 5.9 and 5.10a, under normal operating conditions the source of initiative is the user in charge of the switch (transitions from state 1 to state 2, and from state 3 to state 4). The light bulb itself initiates the free-will transition of the failure mechanism (state 3 to state 5).

Classification of state transitions – 2

Transitions are also classified according to the meaning assigned to the states they link.

- *Operational transitions* originate and terminate in operational states.
- *Non-operational transitions* are processes which originate in a non-operational state, terminate in a non-operational state, or both.

For VERSIONS C and E of the light bulb example, the results of these state transition classifications are given in Figure 5.11 and 5.12, respectively. The 'COMMENTS' column shows the classes to which each transition belongs.

Classification of state transitions – 3

Depending on which of the two components of the closed world changes state, we have a further criterion of classification:

- *environment-active* transitions and
- *target-active* transitions,

This last criterion applies to all processes except those of state maintenance, since, in this latter case, neither component of the closed world system changes state.

The state transitions, together with the designation of their classes, are shown in Figures 5.13 and 5.14 for VERSIONS C and E, respectively.

Concluding remarks on classification of state transitions

Transitions which are both environment-active and target-active are not compatible with the concept of a responsive system, defined here as one where

	CURRENT STATE	NEXT STATE	COMMENTS
1	OFF,ON	OFF,OFF	state maintenance, operational
2	ON,OFF	ON,ON	transient, operational, target active
3	ON,ON	ON,ON	state maintenance, operational
4	OFF,ON	OFF,OFF	transient, operational, target active

Figure 5.11:
Version C, Part 2 - The forced transition table of the working light bulb and switch

	CURRENT STATE	NEXT STATE	COMMENTS
1	OFF,ON	OFF,OFF	state maintenance, operational
2	ON,OFF	ON,ON	transient, operational, target active
3	ON,ON	ON,ON	state maintenance, operational
4	OFF,ON	OFF,OFF	transient, operational, target active
5	ON,FAIL	ON,FAIL	state maintenance, non-operational
6	OFF,FAIL	OFF,FAIL	state maintenance, non-operational

Figure 5.12
Version E, Part 2: The forced transition table of failure-prone light bulb and switch

	CURRENT STATE	NEXT STATE	COMMENTS
1	OFF,OFF	ON,OFF	transient, operational, environment active
2	OFF,OFF	OFF,ON	transient, non-operational, target active
3	ON,ON	OFF,ON	transient, operational, environment active
4	ON,ON	ON,OFF	transient, non-operational, target active

Figure 5.13:
Version C, Part 3 - The free-will transition table of the working light bulb and switch

	CURRENT STATE	NEXT STATE	COMMENTS
1	OFF,OFF	ON,OFF	transient, operational, environment active
2	OFF,OFF	OFF,FAIL	transient, non-operational, target active
3	OFF,OFF	OFF,ON	transient, non-operational, target active
4	ON,ON	OFF,ON	transient, operational, environment active
5	ON,ON	ON,FAIL	transient, non-operational, target active
6	ON,ON	ON,OFF	transient, non-operational, target active

Figure 5.14:
Version E, Part 3: Free-will transition table of failure-prone light bulb and switch

only one component of the closed world can change at a time, by reacting to the change of the other. This set comprises part of the difference between the theoretical and actual state transition capacity of the FSP, and lies outside the specification for the chosen modelling method.

5.4.4 Summarizing the classification of states and state transitions for measurement

The measures of the FSP (Figures 5.15 to 5.18) arise naturally from the state transition model, and from the classification of states and state transitions shown in the tables of this chapter. They are derived by counting, and applying the equations of Section 5.4.3.

For the purpose of measurement, the closed world FSP specification is expressed in three parts. These may be compiled using the state transition diagram, but the diagram itself is only an ancillary part of the specification.

- Part 1 is a *list of the states* of the two components of the closed world.
- Part 2 is a *forced transition table*. This table contains all the states of the closed world and a 'next state' defining the transition. Each of these transitions is then classified according to three criteria:
    - (i)   state-maintenance or transient;
    - (ii)   operational or non-operational;
    - (iii) environment-active or target-active (as already mentioned, this last does not apply to state maintenance transitions).

- Part 3 is a similar *free-will transition table*, showing classification as:
    - (i)   operational or non-operational,
    - (ii)   environment or target active.

Parts 2 and 3 are presented separately only for convenience, using the distinction between forced and free-will transitions. Each of the other criteria could be used for a first partitioning. The level 1 measures themselves will consist of separate counts of sufficient categories to enable the rest to be deduced as indirect measures.

Analysis of the classes of states in Part 1 and transitions in Parts 2 and 3 leads to direct measures on the specification. For VERSIONS C and E of our example, these are shown in Figures 5.15, 5.16 and 5.17, respectively. These tables also contain some indirect measures which arise from the classification. These tables facilitate direct comparison between the two versions of the example on the basis of the measures. Additionally, and for completeness, Figure 5.18 provides some 'general measures' which apply to all systems of given size under the given operational assumptions.

Direct measures			Indirect measures		
measures for example version:	C	E	measures for example version	C	E
environment state count	2	2	state capacity	4	6
target state count	2	3	non-operational state count	0	2
operational state count	4	4	transient state count	2	2
stable state count	2	4	operational transient state count	2	2
operational stable state count	2	2	non-operational stable states, non-recoverable error measure	0	2
			non-operational transient states, the redundant states	0	0
			utilization of states	1	.66

Figure 5.15:
Forced state transition measures for Part 1 of the light bulb specifications

Direct measures			Indirect measures		
measures for example version:	C	E	measures for example version	C	E
environment active operational forced transition count	0	0	operational forced transition count	4	4
environment active non-operational transition count	0	*	forced transition count	4	6
			state maintenance transition count	2	4
			operational state maintenance transition count	2	2
			non-operational maintenance transition count, non-recoverable error states	0	2
			non-operational forced transition count, redundant states	0	0
			target-active operational transition count	2	2
			target-active non-operational transition count	0	0

*derivable indirectly

Figure 5.16: State measures for Part 2 of the light bulb specifications

Direct measures			Indirect measures		
measures for example version:	C	E	measures for example version:	C	E
operational free-will transition count	2	2	free-will transition count	4	6
environment-active operational free-will transition count	2	2	non-operational free-will transition count	2	4
			etc.		

Figure 5.17:
Free-will state transition measures for Part 3 of the light bulb specifications

Indirect measures		
measures for example version:	C	E
state transition capacity	16	36
state transition count	8	12

Figure 5.18: General measures for the light bulb specifications

The direct and indirect measures together amount to the 'data base' of the object-oriented measures. The specifier may develop from these further (objective) indirect measures, should these be required, and devise over them (subjective) utility measure, to reflect the demands of a given application and quality procedure, thus completing the measurement scheme for the FSP.

5.5 Summary

Models of products and processes suggest their own measures, and lead naturally to the design of measurement schemes. Products can be specified independently of the processes which bring about their referents; thus, it is relatively simple to define for their specification a comprehensive measurement scheme, including object-oriented and utility measures. The basis of such a measurement scheme for products and product species is the black-box model, complemented by property measures. Measures for the design of products are derived from the structural model of the product, defined over specification measures of the atomic products to which the structure resolves.

Processes are specified with reference to their products, and hence their measures are composable from the measures of their input and output product specifications. Process designs are structures over atomic processes, and their measures are corresponding compositions. Because of the relative complexity of process specifications arising from their context-dependence, only general principles could be discussed, but it has not been found feasible to construct for them a general-purpose measurement scheme.

Instead, in this chapter the measurement of specifications of finite state processes was presented in some depth. Finite state processes, coupled to product/process models, offer a convenient model for representing the life history of complete artefact families in operation, and have been used in the specification and design of switching systems of many kinds. The state model used here treats the system and its environment symmetrically: they are both parts of a complete closed-world system. The measurement scheme relies on the classification, and the measures are counts of the population of classes. By classifying the states and state transitions to which finite state processes are subject, it becomes possible to assure specifications and designs against some important classes of errors, such as incompleteness of the description of the desired black-box behaviour, and omission of non-operational conditions. The classification forces the designer to identify those parts of the closed-world system which are open to random behaviour, such as environmental intervention, hardware failure, or free-will decisions by human operators within the system, and to take explicit account of the behaviour of this type of 'system component'. The number of members in each class of state and each class of state transition gives an objective characterization of the specification of finite state processes, and the measures can be used in constructing the utility measures of the measurement scheme.

The finite state systems presented in this chapter were, by necessity, of very limited size. However, the measurement scheme shown has been used on several industrial examples. Among these have been cases arising from telecommunications, and the authors are obliged to BT for supporting part of the relevant research (e.g.[2], [3], and [5]).

The first such example demonstrated the use of the closed-world model in specifying a simple communication system, isolating a single user as the 'environment' component of the closed world. The target system then embraced all the rest of the communication system, including other users. This example also served to show the procedure of constructing the system specification, starting from the 'operational' state transition model, developing this into state-complete and transition-complete models for quality, and compiling the relevant measures.

Subsequent examples developed in two directions:

- the simple system of the first example was extended into a larger, more realistic communication system, and
- the closed-world model was used to support design, as well as specification, through stage-by-stage refinement.

Examining the finite state process from the viewpoint of their implementation at the more detailed 'physical' level, states of the process encode sets of physical properties measurable within the domain of classical metrology. Thus, the measurement scheme of the FSP could be extended downwards in the system hierarchy, and the measures integrated into the established measurement schemes of other domains of science and technology. The FSP measures could also be incorporated into the system structure, thus extending the measurement scheme upwards, to form measures of the higher levels of the system hierarchy.

5.6 References

1 Juran J M, Gryna F M (1983): "Quality planning and analysis". McGraw Hill.

2 Myers M: (1990) "Quality assurance of specification and design of software." Ph.D dissertation, South Bank University

3 Kaposi A A, Myers M (1990): "Quality assuring specification and design." Software Engineering Journal, Vol 5, No 1 pp 11-26. IEE

4 Harel D (1987): "Statecharts: A visual formalism for complex systems." Science of Computer Programming 8, pp 231-274, North Holland.

5 Kaposi A A (1990): "Classification". Task 4, BTRL Project SE00361: "Framework for Network Modelling".

6 CONSTRUCTING A MEASUREMENT SCHEME

6.1 Introduction

Previously we demonstrated model-based measurement on simple examples, and discussed the notion of a model-based measurement scheme and its role in the validation of measures. We showed that a model-based measurement scheme is already implicit in mature disciplines whose models, theories and metrology are well established. The general applicability of model-based measurement indicates that one may also devise model-based measurement schemes for new fields of application. where such foundations are lacking.

This chapter seeks to demonstrate the feasibility of model-based measurement in an important and difficult application area, where the referent is abstract and controversial, and the use of measurement is not yet established. By following the case study demonstration with care, the reader will gain insight into the initial construction of a model-based measurement scheme, and will see from the discussion how such a scheme may be developed and applied in other fields. The demonstration serves as a preparatory step towards developing practical measures for specifications.

6.2 Devising the case study demonstration

Choosing specifications as the application domain

Almost any application area could have been selected for this case study demonstration. We chose the domain of specifications for three reasons: its industrial interest, its difficulty, and its wide applicability.

The need for improved specifications has led to industrial demands for 'specification metrics', and for methods of objective appraisal of the 'quality' of specifications. The requirement was expressed in vague general terms, without defining what constitutes quality in this context, and without making explicit the attributes to be measured. The novelty of industrial use of formal specification languages, the diversity of formal and semi-formal specification notations, and the unfamiliarity of characterizing such specifications by measurement, contributed to the difficulty of the subject area.

The measures of the scheme were considered arbitrary, and the validation of meaningfulness was expressly excluded from the aims of the case study demonstration. Wide applicability was of central interest. We considered that if it were possible to construct a model-based measurement scheme for specifications expressed in some 'general purpose' specification language, then model-based measurement would be applicable as widely as would the specification language itself.

Choosing the specification medium

Characterization of specification and design by measurement calls for a formal model. The specifier/designer can select any language, as long as it can manifest the key attributes of the referent. A variety of powerful models have been developed in the specialist disciplines of science and engineering. Where the problem can be solved within the boundaries of a specialist discipline, the most elegant representation is usually one which is tailor-made for the given referent and the problem in hand. The difficulty is that the field of application of each kind of model is restricted, either by some inherent limitation of the model itself, or simply by tradition, and by the cultural divide between disciplines.

For these reasons, there is a strong case for selecting a 'wide-spectrum' language with broad applicability and a large community of users. The natural language (English, say) is the obvious choice, and it is the most widely used medium of system specification at present. It is a matter of concern that in current practice sophisticated computer-based systems are specified in the natural language informally, albeit with some formally defined parts, the system specification amounting to many volumes of language text. Such a specification is open to error, omission, misinterpretation, and misunderstanding between client and supplier and among members of the multidisciplinary team of designers and implementers. The outcome is delay, loss of system performance, as well as poor reliability, poor safety, and indifferent general quality. Examples are numerous, and widely reported in the press.

To remedy this situation, several 'general purpose' languages have been developed by large industrial concerns for their own use, or else by academic institutions, sometimes with the backing of such concerns. These languages are formal or semi-formal systems with well-defined syntax and semantics. The syntax of a specification language may be textual, with complementary mathematical symbols, or graphical, again with textual and mathematical complements. The syntax offers 'terminal symbols' which stand for the smallest meaningful parts of the language (words), and there are 'grammatical rules' for assembling these into larger entities of the language (sentences, sequences of sentences, and complete documents), with semantics providing rules for interpreting the meaning of syntactic parts. Although they are referred to as 'specification languages', most can describe both the specification and the design of the referent. For some safety-critical applications the use of such languages is already established, and is prescribed by military and civilian standards. Formal and semi-formal languages are increasingly being adopted in other application areas. Since these languages claim 'generality', in the sense of power to represent referent systems drawn from a wide application domain, we base our *model-based measurement strategy* on them: whatever such a well-defined language can model, we can then measure.

Particularly valuable for the system designer are those formal languages which go beyond symbols and rules of description, having in addition, a *formal reasoning system* associated with them, employing suitable subsets of mathematics. The reasoning system supports manipulation while preserving meaning, and facilitates drawing of deduction, for example, to confirm consistency between two documents, one describing the specification of the referent and the other the proposed design. Since our particular interest is model-based measurement in support of specification and design, we build our strategy on these kinds of languages, although we show that the strategy is applicable to other, less formal languages.

Outline of model-based measurement of specifications

The model-based measurement process is shown again schematically in Figure 6.1a. In our demonstration study, modelling becomes a composite process. The initial model (the natural language text, say) is first expressed in the specification language, and then the resulting formal or semi-formal language entity is itself treated as a referent and is subjected to model-based measurement. The whole process, from original referent to measures, is shown in Figure 6.1b. The figure describes in extended form the same process as Figure 6.1a, but it has been rotated by 90° for ease of representation on the page. Validation arrows have been omitted in this and subsequent figures, in the interests of simplicity.

Figure 6.1b points to some possible uses of model-based measurement of which the first is directed at the characterization of the original referent:

- Assume that the specification is a formal (or semi-formal) language entity. If this is a valid representation of the referent, and if the model of the language entity preserves the relevant attributes of the specification, then at least *some* of the measures of the model will be

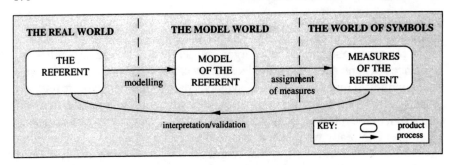

Figure 6.1a: Outline of model-based measurement

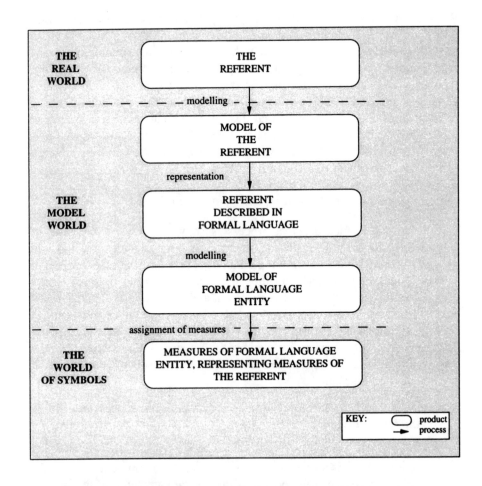

Figure 6.1b: A model-based measurement strategy, using a formal language

representative of the *referent,* and these will be suitable for entry into the model-based measurement scheme for specifications.

Some of the measures of the model may capture the attributes of the original referent; other measures may reflect the features of the specification language itself. The latter may be irrelevant in characterizing the original referent, but they will be important in deciding which specification language to choose in a given situation. This leads to the second use of model-based measurement:

- Further assume that the specification is a formal (or semi-formal) language entity which is a valid representation of the referent. At least *some* of its measures will be representative of the characteristics of the *specification language*, and these will be fit for entry into the model-based measurement scheme for specifications.

Validation of measures and measurement schemes

Validation of the feasibility of constructing a model-based measurement scheme is distinct from the validation of the measures themselves. It may be that a measure is included in the measurement scheme with the intention of characterizing the referent, but proves not to stand the test of validation in problem solving practice.

- The measure may properly represent some attribute of the referent, but the attribute may not be one of the key features of the individual specification problem.

- The measure may have been unwisely chosen for the given case, but may prove to be a valid and useful measure for some problem in the future, and is worth preserving for future reference. For the present case, the measure would need to be modified, either by combining it with other extant measures of the scheme, or by adding it as a new measure at the next phase of development of the measurement scheme. The measure validation has failed, but the measurement scheme has contributed to the state of knowledge in the *problem domain.* If the measure emerges from a well-constructed measurement scheme, it may also reveal some new emergent property of the *referent* whose importance was not recognized by the problem solver at the outset.

A valid measure of the *specification* may not be a valid measure of the *referent,* nor would it be for a referent of any future specification problem. There can be at least two reasons for this (see Figure 6.1b):

- The specification may have misrepresented the referent. Here the fault lies with the modeller, or with the inadequacy of the specification language. The designer of the measurement scheme can do nothing about it; however, the measurement scheme is a useful aid in diagnosing the error.

- The measures, devised for characterizing the referent, may show incidental features of the *language,* rather than relevant features of the original *referent.* Measure validation fails, but the effort is not wasted if the measures reveal important features of the specification medium.

(An absurd example shows that this is not always the case. Assume that a measure of the specification of a motorcar, given in some textual language, counts the number of occurrences of the letter 'p' in the text. This is unlikely to have any connection with any of the attributes of the car, or show any interesting feature of the specification language.)

If features of the referent and the specification medium are not distinct but are combined within the same measure, this might yield misleading conclusions. We indicate in the demonstration study some of the ways in which model-based measurement assists in guarding against interpretation errors of this kind.

Formal languages in model-based measurement of products

Consider now that the referent is a product. We know that the *specification* of a product is its black-box model, qualified by an appropriate collection of variable measures, whereas the *design* of the product is a structure over the specifications of the product's components. To be included in our model-based measurement scheme, it is useful if the language can describe both the specification and the design. Equation 6.1 serves as a summary and reminder of the relationship between system specification and design:

$$M_B(P) \subseteq M_B(M_S(P)) \dots\dots\dots\dots\dots\dots\dots\dots\dots\dots\dots\text{Eq 6.1}$$

where **P** is the referent product,

$M_B(P) = (\textbf{Prop}, \textbf{R}_{Prop})$ is the black-box model of **P**, becoming the prescriptive specification by the assignment of parameter values to each property variable $\textbf{prop}_i \in \textbf{Prop}$,

$M_S(P) = (\textbf{Comp}, \textbf{R}_{Comp})$ is the structural model of **P**, which thus becomes the design of **P** by assignment of a specification to each component $\textbf{comp}_j \in \textbf{Comp}$ within the structure $\textbf{R}_{Comp}$,

$M_B(M_S(P))$ is the black-box model of the design of **P**: the descriptive specification of the referent,

and $M_B(P) = M_B(M_S(P))$ is the limiting case of equation 6.1, where the design has exactly the same property set as the specification.

Assume now that the prescriptive specification $M_B(P)$ of the product P is given in a formal language, and that the structure of the design (R_{Comp} of P) and the specification of the component set (**Comp**) are also described in the same language. Then, if the language has a suitable formal reasoning system, one may first deduce the descriptive specification of the product from its structural model, and then verify the correctness of the product design, proving the consistency between the descriptive and prescriptive specification. If the formal language can reason in terms of variables, then it is possible to prescribe and describe in the formal language the specification and design of a whole product *species*.

Each stage of the modelling process of Figure 6.1b treats the previous stage as its referent. If the purpose is to characterize the specification of the product – the original referent – its attributes must be preserved through the modelling chain, but other features of the specification language can be suppressed. The simplest guard against measuring trivialities of the language rather than the message expressed in it is to assure that the models of the measurement scheme correspond to meaningful syntactic entities which refer to the product.

We know that the measurement scheme must discriminate between objective attributes and subjective judgements about them. At the start of developing a model-based measurement scheme, a collection of object-oriented measures must be assembled for the former, leaving the user free to devise measures for the latter. Then, given an appropriate collection of object-oriented measures, subjective utility measures can always be added at subsequent stages of the development of the measurement scheme.

Formal languages in model-based measurement of processes

We defined an actual process as a function $q(\delta t)$ which maps an input product $P_{in}(t_{in})$ into an output $P_{out}(t_{out})$. δt is the only time parameter of the process, to be set by the specifier, and to be met by the design. We do not refer to it again, but assume that any formal language within our strategy can acknowledge the finite duration of processes, at least implicitly.

A formal language may be 'procedural', representing the referent process $q(\delta t)$ in a simulation model. Then applying to the simulator the input $P_{in}(t_{in})$ would yield the output $P_{out}(t_{out})$. The specification would normally be wider than just an actual process for an individual product, referring to product species or to a whole product genus, depending on the degree of instantiation of property variables, as illustrated in Figure 4.13. For all but the simplest processes, complete enumeration of all $P_{out\,i}(t_{out\,i}) \in Q_{out}$, $P_{in\,i}(t_{in\,i}) \in Q_{in}$, will be unfeasible, or at least lengthy and laborious, with the result difficult to comprehend, especially if the process has memory and Q_{out} must keep track of the state of the process.

As we know from Chapter 4, the alternative is to specify the process indirectly, through its product specifications. A 'declarative' language describes the process by its input and output products, rather than give a procedure for obtaining one from the other. The correctness of the process specification can be proven by calling on the reasoning system of the language. In many cases the product specification is itself implicit, defined through the design of the product, and if the process has memory, the declarative specification would give the state of the products before and after each instantiation of the process.

Requirements of languages for the measurement strategy

The requirement that a language for expressing specification and design be a *formal* system is basic to the notion of derivability and proof of correctness. The formalism seeks to assure that statements in the language are *precise* and *unambiguous*. The language should express all the features of the specification/design of the original referent, and must suit the environment in which it is to be used. The particular formal system chosen to model the product or process must be *expressive*, in the sense of being able to represent a variety of models, and *general*, capable of modelling a wide variety of referents, thus justifying the effort invested in its learning and implementation. The language definition must be *complete*, extending to all aspects of syntax and semantics, and it is desirable that the language should be *concise*, using the smallest set of symbols and rules consistent with other requirements.

The attributes just described refer to the language *medium*, as well as the *message*. In identifying further desirable attributes, one must bear in mind that the main purpose of a language is communication. As far as possible, the language must be *natural* and *easy to use*, and statements in the language should be *easy to understand* by the intended recipient.

It is highly desirable that a formal language be *executable*. This allows the declarative language to be treated as a programming language, and used effectively in writing large specifications, describing the design of their parts, testing and debugging specifications, and assuring consistency between specification and design. Executability assists in validation of the specification against the client's recognized and hidden requirement, such as by 'animating' the specification to demonstrate characteristics of the proposed system, trying out options, and experimenting with changes. The executable specification is a powerful tool of requirement capture before, during, and after the development process.

The need to express specifications in a formal language is gaining reluctant acceptance in industry, and many formal languages exist and are being created. To date there are no established criteria for matching language to problem, or for assessing the effectiveness of these languages.

6.3 Measurement strategy options

Our object is to develop a 'general' strategy for model-based measurement of specifications and designs in accord with the outline in Figure 6.1b. This purpose would be poorly served by finding measures of specifications expressed in a single specification language, especially since there is no consensus as to the 'best' among the many formal and semi-formal languages on offer. Instead, we propose a strategy for measurement of process models expressed in a *variety* of specification languages. The measures may then serve in objective appraisal and subjective judgements of the specifications themselves, the processes they model, and the languages in which they are expressed.

We now explore the options for implementing such a strategy.

6.3.1 Single-language deductive strategy

Assume now that a suitable formal language has been found for expressing the model of the specification/design of a class of systems. Assume also that measures can be defined for all meaningful syntactic parts and constructs of the language, such that the measures remain invariant under the rules of manipulation, and that the model also meets the criteria of measurability drawn up in Chapter 3. By establishing a general 'measurement scheme' for the language as a whole, one can apply measurement to any syntactically correct statement in the language. Thus, there is no need to invent measures for individual specifications or designs: one may *deduce* from the general measurement scheme of the language the measures which characterize any specification/design expressed as a legal statement in that language.

We refer to this as the 'single-language deductive strategy' of model-based measurement.

Measurement in the single-language deductive strategy

Measurement of the characteristics of an object such as a specification will not produce a single number, but a whole *set* of numbers or other symbols, or even sets of sets. Many factors will contribute to characterizing different features of such a referent, and one would attempt to assess these independently.

Let us first restrict our attention to a single referent, **S**, and let the required attributes of **S** be described, in the first instance, in natural English. This description is then expressed more formally in the language, **L**1, for which a unique measurement scheme exists, so that the specification is characterized by deducing its measures. If the design task is critical then a 'diverse' specification method may be used to enhance reliability. Here the natural language description is given to a number of specifiers, working independently, and comparing the results they produce. Such a situation is shown in Figure 6.2.

Experience shows that measures of two valid specifications, derived from the same natural language description of the same referent, may differ considerably. The difference is attributable to various reasons, such as:

- specifiers have different appreciations of the referent, and put different interpretations on the natural language description;
- using previous experience with similar systems, specifiers implement different characteristics which are not explicitly required, but which they judge to be desirable;
- there is a difference in the rigour applied to the specification, and the detail to which the design is resolved;
- there are stylistic differences which allow different valid statements within the rules of the language.

Analysis of the measures of an individual specification, and of versions, will aid the development of quality assurance procedures, and indicate future guidelines to specifiers.

Assuming all the factors just referred to are monitored and guidelines for constructing specifications are in place to minimize variations of specifications of a given system, then this same single-language deductive strategy allows direct comparison of specifications of different systems in the language L1, as in Figure 6.3.

6.3.2 Multiple-language strategies

The strategy just described serves the needs of the user community of one particular specification language. This is not adequate for our purpose at the present state of the art. Many different formal and semi-formal specification languages are in use and under development. The various languages compete for the favours of the potential user community; hence their measurement-based appraisal is increasingly important.

It would take a great deal of effort to develop a comprehensive model-based measurement scheme for each formal language individually, and even then, there would be no basis for making inter-language comparisons. Instead, a measurement scheme is required which is not confined to a single language, but which transgresses language barriers, so as to allow comparison between them, and between specifications/designs written in different languages. Such a 'multiple-language' measurement scheme could then be used to facilitate the setting of standards applicable to specifications/designs for all languages in the scheme.

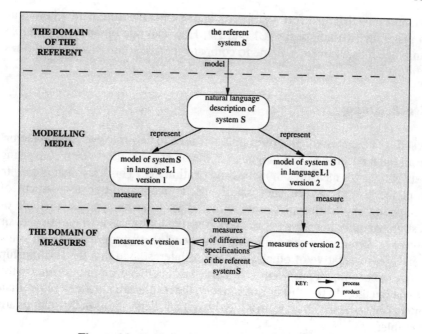

Figure 6.2: Single language measurement process

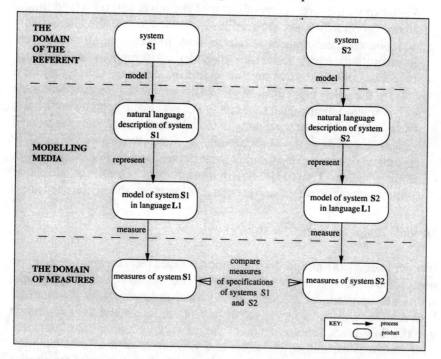

Figure 6.3: Comparing specifications of different systems described in the same language.

Assume now that no effort was spared, and distinct measurement schemes are in place for two languages, **L1** and **L2**. How can one compose a multiple-language measurement scheme to compare specifications written in the two languages?

Benchmarking

A series of benchmark programs might be used to define a 'conversion factor' from the measures of one language to the other, as is used when converting between length measures in inches and centimetres. This approach is easy to implement, and can be successfully used in comparative assessment of competing application programs with respect to some specific attribute of resource requirement or performance. Benchmarking has no theoretical basis however. It relies on the notion that the benchmark programs are, in some sense, representative of other programs of interest, and that the relationship between benchmark programs is preserved in all other programs written in the two languages. In the absence of a theory, the results are *numbers* rather than (model-based) *measures*, from which no reliable derivations or deductions are possible.

When expressing reservations about the benchmark approach (and similar quantification schemes), we readily acknowledge that the alternative – model-based measurement – has its own limitations. All models are simplifications which reflect the modeller's assumptions and although subjective utility measures are built up from objective measures, they must embody the value judgement of the measurer; the implementation of the measurement scheme must take account of pragmatic appraisal of what is feasible. Nevertheless, the quality of model-based measurement is assured because the measures are repeatable, all subjective characteristics are explicit, and are traceable to small, well-defined areas within a framework of the scheme. The model-based measurement scheme is subject to continuous reassessment, and is open to refinement, should the need arise.

Multiple language deductive strategy

In order to devise a multiple-language model-based measurement scheme, we rely on the notion of 'levels' of languages. This notion is widely applicable, and could be used, for example, in electrical engineering, to discriminate between the 'high-level language' of twoport networks and their resolution into the 'low-level' representation of the network as a structure of uniports, with resistors, capacitors, inductors and generators as low-level language elements. Such circuits can be resolved further, to the language of the physics of the mechanisms responsible for resistance, capacitance, etc.

'Level' is a concept well understood in the field of programming languages (see e.g [1]). Algorithms are implemented by statements in machine code: the lowest level programming language (ignoring microcode). *Higher-level* languages, such as Pascal, Fortran, Cobol or C, group together sets of machine instructions into a single statement, thereby creating features specific to the given language. The fewer programming statements needed to implement a given algorithm, the *higher* the language level, measured on a vertical scale of descending order between the human/machine interface and the machine hardware. By their conciseness, the higher-level languages ease the task of the programmer. To execute them, Pascal, Fortran, Cobol, C and other higher-level language programs are all converted into the lower-level language of a common machine code: a neutral representation of the program. The conversion is implemented by a *language translator,* which may be either a compiler or an interpreter ([2]). The mapping between the original and the translation can be reversible, provided that sufficient information is preserved about implementing the translation.

The notion of language levels is readily applicable to specification languages. A language which uses more statements to express a given specification is a lower-level language than the one which uses fewer statements to express the same specification. If one can identify a suitable lower-level formal language into which other, higher-level formal languages could be translated, then the lower-level language may be used as a neutral 'reference language' for implementing a simple and effective unifying measurement strategy for a whole family of higher-level formal languages. Only the one model-based measurement scheme would need to be developed: for the reference language alone. If translation of all the other languages is incorporated into a model-based measurement process, then we have obtained a 'multiple-language' model-based measurement scheme. Legal statements in all other languages of the family could then be indirectly measured in the translation. Such a multiple-language measurement scheme can be used in a single-language environment to characterize specifications written in any given language of the family but can also serve in a multilingual situation, measuring several formal languages simultaneously, to compare them objectively.

We refer to this as the 'multiple-language deductive strategy' of model-based measurement. Figures 6.4 to 6.9 show the strategy at work.

Using multiple-language measurement in a single-language environment

Figure 6.4 recreates the case of Figure 6.2 when a natural language description of a referent system **S** gives rise to two different specification versions of **S**, written in the same language. In this case the formal language is **L**1, for which no suitable measurement scheme is available. Rather than devising a whole

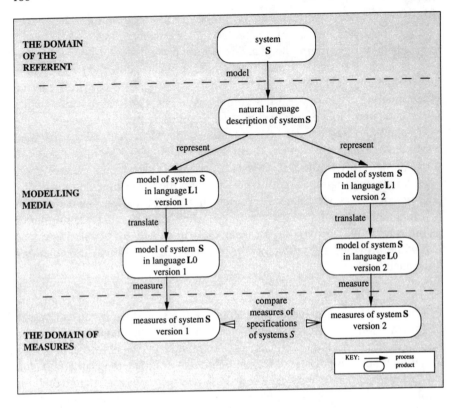

Figure 6.4: Single-language model-based measurement of two specification versions of system S in language L1, and implemented through use of the reference language L0

new measurement scheme for **L1**, model-based measurement is achieved indirectly, by translating both versions from **L1** into a suitable reference language **L0**, for which a model-based measurement scheme has already been established. Once a translation method exists, there is no need for the **L1** user community to learn any detail of the implementation of the measurement scheme, and hence no knowledge of the reference language is required. Model-based measurement will thus facilitate a transfer of measurement technology from language **L0** to language **L1**.

The strategy is also effective when the specification of two systems, **S1** and **S2**, are to be compared (Figure 6.3). Here again, the measurement of specifications expressed in **L1** is can be done indirectly, through translation from language **L1** into **L0** (Figure 6.5).

Figure 6.6 shows yet another use of model-based measurement: as the means of comparing different designs of the same system, to appraise design alternatives.

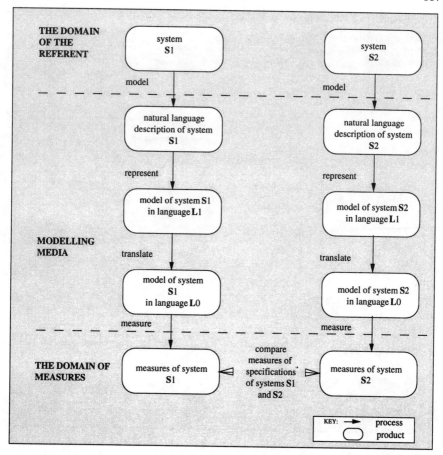

Figure 6.5: Single-language model-based measurement of specifications of two systems in the language L1, implemented through use of the reference language L0

The multiple-language measurement strategy, based on L0, is implementable in principle for the **L1** environment, as long as the concepts expressed in **L1** can also be expressed in **L0**. The strategy is also implementable in practice, as long as a suitable translator from **L1** to **L0** exists.

Multiple-language measurement in a multiple-language environment

Assume now that the concepts of the specification languages **L1** and **L2** are expressible in the 'reference language' **L0**, which means that they are higher-level languages than **L0**, and both **L1** and **L2** can, in principle, be translated into **L0**. Assume also that translators are available, and that a measurement scheme for **L0** already exists. The measurement of specifications and designs,

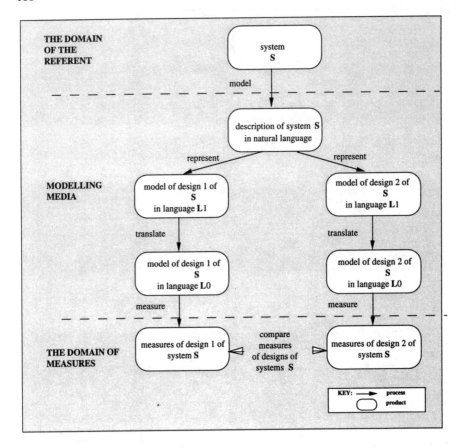

Figure 6.6: Comparing different designs created to the same specification,
implemented in the same Higher-level language

written in **L1** and **L2**, is then tackled as shown in Figure 6.7. The measurement
scheme can be readily extended to further higher-level languages **L3, L4**, etc.,
provided that the same conditions also apply.

Other uses of the multiple-language measurement

Assuming valid translation, further possibilities arise from multiple-language
measurement as refinements or developments of the scheme of Figure 6.7.
We only mention specifically, three of these.

(1)   In the scheme of Figure 6.8, L0 serves two functions. It is the reference
      language of model-based measurement into which **L1** is translated as
      before, and it is also used as a specification language in its own right.

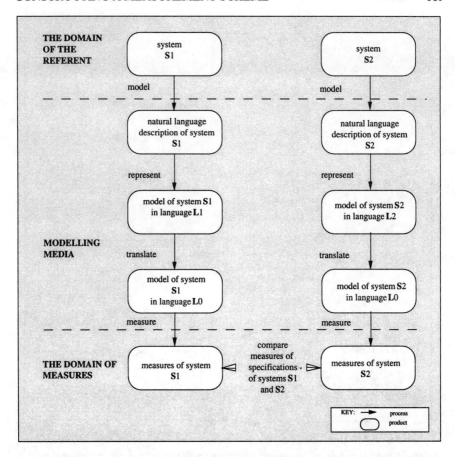

Figure 6.7: Characterizing specifications of different referents, represented in different languages

Measurement can be used to compare two versions of the specification of a given system **S**, or else for comparing the performance of two specification languages, the higher-level language **L1** and the lower-level language **L0**.

(2) Using measurement to compare two specifications of the same system, expressed in different higher-level languages in support of diverse specifications, and also to facilitate the comparative appraisal of two languages **L1** and **L2** (Figure 6.9).

(3) Analyzing specifications and constructing profiles of referents.

Direct measures of a given specification could be combined in an appropriately constructed model to give revealing indirect measures or object-oriented characterizations of the referent, as shown in Figure 3.6.

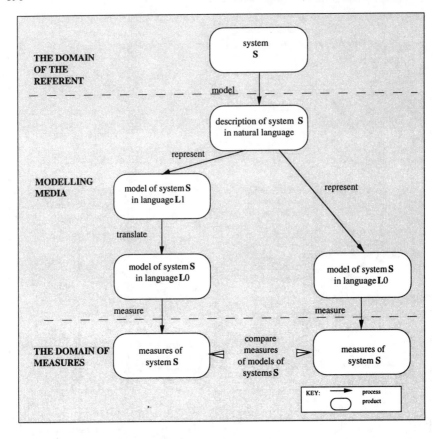

Figure 6.8: Comparing a Higher-level language specification of a given system with a specification represented in the reference-language

The measures may indicate the *class* of the referent; for instance, one would expect that information retrieval systems would have different attributes from, say, real-time software systems, and measures of their specifications should reflect this difference. The measures may also allow insight into the *size* and *resource requirements* of a given class of referent, thus providing early guidance for the time scale and costing of projects, and assist procurers in the comparative evaluation of competing proposals. In both cases, deviation from expected object-oriented measurement profiles would be of value in quality monitoring. Characterization of modules of mixed systems could suggest novel groupings, leading to new design possibilities.

The exploration of such uses of model-based measurement are touched on in Chapter 10. They would open up new fields of research into the specification and design of systems.

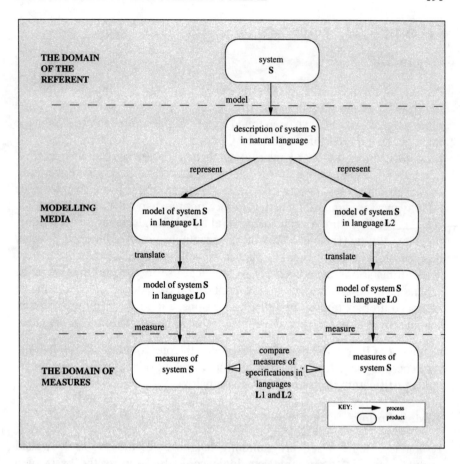

Figure 6.9: Comparing two different high-level language representations of a given specification/design

6.3.3 Choice of a measurement strategy

The multiple-language deductive strategy has clear advantages over a single-language strategy or a benchmarking strategy. It meets all the measurement demands of the single-language strategy and, unlike benchmarking, it gives a sound basis of comparison. If the referent language is well chosen, the multiple-language strategy has the flexibility of incorporating an increasing range of formal languages in one extensive measurement scheme.

To show the feasibility of this strategy, we have to:

- identify an appropriate reference language L0;
- create a model-based measurement scheme for that language;
- show how other formal languages translate into the reference language within a model-based measurement scheme.

6.4 A unifying reference language

Why Prolog?

Prolog is proposed as a suitable reference language for implementing our multiple-language measurement strategy. The choice is not unique. Some other language might have been found equally suitable, and another language may become preferable in future. Our purpose was to demonstrate the feasibility of model-based measurement, and Prolog served this purpose admirably. Experience has shown that Prolog manifests the requisite attributes of formal languages listed earlier in this chapter, and it has also proven its effectiveness in industrial practice. For a comprehensive list of examples see e.g. ([3]). The literature reveals that several formal languages have been translated into Prolog ([4]) so as to provide an executable model of the specification/ design, with known advantages, such as testability, animation and validation. These translations have been put to practical use in our multiple-language scheme. Prolog's well-documented application experience offers a mine of demonstration examples. For all these reasons, a model-based measurement scheme for Prolog has been developed, as is demonstrated in later chapters.

Chapter 7 of Part 2 of the book is devoted to the presentation of Prolog as the reference language for such a strategy, but a brief introduction to the language, and a justification of its choice, is given below.

What is Prolog?

Prolog is a logic based programming language. In the 1970's Kowalski showed that first order predicate logic could be used as the basis of a declarative programming language ([5]). Such a language is particularly suited for expressing specifications because it concentrates on *what* is required, rather than on *how* the requirement is to be met. Combined with Robinson's resolution method ([6]) and his unification algorithm, Warren ([7]) and others ([8]) were able to write efficient compilers, and the Prolog programming language was born.

Prolog programs take the form of *facts* and *rules* (also referred to as *assertions* and *implications*) which together form a *data base*. A query to the data base is a new fact which can give rise to three sorts of outcome (see e.g. [9]):

- The new fact is proven *false* if it is inconsistent with the original data base. Then the language translator returns the simple message 'No'.

- The new fact is proven *true* if it is unconditionally consistent with the original data base. In this case the language translator returns the response 'Yes'.

- The new fact is proven *conditionally true*. In this case the language translator returns the values of the variable domain for which the fact is true.

Facts take a form such as:

>   metal_ion(barium).
>   acid_ion(sulphate).
>   insoluble(barium,sulphate).

Rules have two parts: a consequent and a condition list. It is in the form: 'consequent if condition_list'. An example is:

>   non_poisonous(A,B):-metal_ion(A),
>   acid_ion(B),
>   insoluble (A,B).

With just this one rule and the two facts in the data base, we can illustrate the three types of outcome to be expected.

- To the query '?non_poisonous(barium,sulphate)', the response is 'Yes', since the queried fact is unconditionally *true*.

- By contrast, the response to the query:

>   '?non_poisonous(sulphate,sulphate)' is 'No',

  a) because there is no fact which declares that sulphate is a metal ion (condition 1), and

  b) because, in Prolog, anything not positively provable is judged false. This last feature, referred to as 'negation as failure', means that there is no way to differentiate between proof of a negative and the inability to prove or disprove the proposition.

- The third case, that of a general query: 'what non-poisonous substances are there?' is entered as: ?non_poisonous(A,B). This would elicit the response:

>   A = barium, B = sulphate.

One can interrogate the data base on further solutions, i.e. other instantiations of A and B, by typing a semi-colon. With such a small data base, this second query would be answered 'No', in other words, there are no other instantiations which fit the known facts.

The facts and rule of the above database, together with a query, form a Prolog *program* which can be read in two ways:

- The *declarative* reading describes the problem.
- The *procedural* reading prescribes the steps to a solution.

The declarative reading of the program above goes like this:

'If a substance is composed of two parts, of which the first is a metal ion and the second is an acidic ion, and if the combination is insoluble, then (regardless of anything else) the substance is non-poisonous.'

The procedural reading, by contrast, is:

"To find out if a substance composed of two parts is non-poisonous, check the following facts: Is the first part a metallic ion? Is the second part an acidic ion? Is the combination insoluble? If all these are true, then the salt is non-poisonous".

The declarative reading of a program is a statement of the problem as a logical structure of related facts. A procedural reading of the same program is a statement of a succession of steps which (should) lead to a solution. This will be so, and the program will be executable, if certain rules are followed, in which case the execution is an *implementation* of the solution, provided that a solution exists.

The danger, of which all Prolog programmers are aware, is that of creating a 'black hole', where the program enters a non-terminating loop, because a fact needed at a certain stage will not be available till later. This means that the program is procedurally incorrect, in the sense that it does not terminate; however, declaratively the program is correct, because it concentrates on the logical statement of the facts and rules, and logic is indifferent to the order in which the statements are made.

Further details and examples of Prolog code are given in later chapters, but the reader who is completely unacquainted with Prolog is advised to consult one of the standard reference books ([10]).

Advantages of Prolog as a reference language for specification

We explore the claim that Prolog contains all the requisite features of a reference language.

- Prolog is a *formal* language, with well-defined syntax, semantics based on the Horn clause subset of first order predicate logic, and derivation rules of logic. Therefore the language is *complete* and *general* enabling a specification written in it to be *precise* and *unambiguous*. There is an associated deduction system. The *model-theoretic semantics* of logic programs assigns the meaning of a logic program to what its statements imply ([11]).

- Since Prolog is *manipulative*, the same program can be presented either in hierarchical or single-level (unfolded) form. The unfolding process is simple, and may be described by an algorithm.

- Prolog is *expressive*, and its use as a formal language in its own right has been advocated (e.g. [12], [13]) and demonstrated. The declarative nature of Prolog facilitates the statement of requirements and facts, without prematurely imposing implementation constraints and details.

- Prolog is *natural* and *easy to use*, as has been extensively demonstrated in school environments and a variety of applications.

- Prolog is *concise*, in the sense of using a frugal syntactic and semantic apparatus. Consequently it is a comparatively low level formal specification language.

- Prolog admits a wide choice of names for procedures and variables, and if these are judiciously chosen then Prolog text is *easy to understand*.

- In its procedural form, Prolog is *executable* and has been widely implemented.

Practical aspects of using Prolog as a reference-level formal language

Several formal languages, other than Prolog, rely on first-order predicate logic, and this common ground supports their translations into Prolog. However, there are practical difficulties in using Prolog as an executable formal language. To resolve these, language designers have introduced elements into the language which complicate the modelling process.

- A serious dilemma, which has long concerned academically minded workers, is the conflict between the requirements of the declarative and procedural use of Prolog. The former needs to maintain rigorous logical foundations; the latter call for the introduction of features which aid communication between the machine and its user, facilitate termination, and enhance efficiency of execution, but which are 'extra-logical'. Without such features Prolog would be of little practical use as a programming medium.

  The measurement scheme aims to characterize specifications rather than program executions; hence it is concerned only with the logic of declarative Prolog, and abstracts away from procedural characteristics.

- Functions and sets are important mathematical structures for describing specifications and designs, and many formal languages make use of them. Prolog cannot handle them within first-order logic, but they are necessary for expressing specifications, and provisions are made for them in the measurement scheme.

- We find in practice that recursion is discouraged or absent in many formal languages, but it is fundamental to Prolog, and its use is included in the measurement scheme.

- Prolog has relatively few language features. This makes the language easy to learn and easy to use, and hence it is a positive characteristic from the viewpoint of novice programmers and occasional users (such as managers, clients and QA staff). However, experienced routine users call for a richer complement of language elements. This has resulted in extensions, enhancements and new dialects of Prolog constantly being developed. Where these are procedural, such as aids to operational efficiency, they pose no measurement problem, since we are only concerned with the declarative parts of the code. However, some include new coding features to ease the task of expressing specifications and designs. This corresponds to altering the level of the language, and is an aspect which may create difficulties with Prolog's use as a stable, unifying, reference language for model-based measurement of specifications. The publication of ISO standards for Prolog ([14]) should alleviate this problem.

6.5 Implementing a multiple-language strategy

Figure 6.10 shows how a multiple-language deductive strategy may be implemented by use of some unifying reference language, L0. The figure describes the process of expressing the specification of referent **S** in any of (n+1) formal languages, and characterizing the specifications by model-based measurement. (Note that in this figure arrows indicate products and blocks indicate processes.)

The specification may be expressed:

- in one of the high-level languages L1 to Ln, translated into the reference language L0;
- directly in L0, which is itself a formal language.

Whichever path is followed, the resulting specification is characterized by a process which models and measures objects in **L0**.

The contents of the rest of the book follow directly from the process described in Figure 6.10.

The starting point of the measurement scheme for Prolog is the logic text. For measurement of existing Prolog programs, the logic text needs to be abstracted from the full Prolog code. **Chapter 7** shows how to strip down a Prolog program to its logic text, how to reveal the structure of the text as

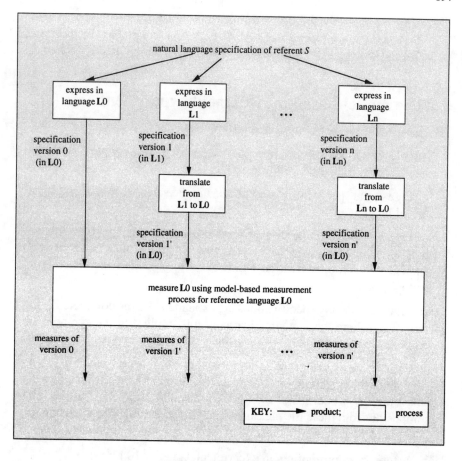

Figure 6.10:
Implementing a multiple-language deductive strategy in the referencd language L0

a hierarchy of logic components, how to represent the data relationships in the text, and how the logic text is used in building specifications.

Chapter 8 introduces a measurement system for the structural relations of the Prolog logic text. This will be a component of measures of design viewed as a structure over specifications of modules.

Chapter 9 is devoted to developing the model and measurement scheme for the data relationships of the Prolog logic text. Data measurement is specifically tailored to the measurement of specifications, where the measures should reflect the logical content of the specification and not the way it, itself, is constructed.

In **Chapter 10**, multiple language model-based measurement is demonstrated through a measurement experiment on a simple specification, tracing various paths through the multiple-language measurement scheme of Figure 6.10, selecting some high-level formal languages (including Z, VDM and Prolog itself), submitting them to Prolog's measurement schemes.

6.6 The model-based measurment experiment

The experiment presented in Chapter 10 proceeds in five stages:

Stage 1 describes the referent to be specified: the simple ventilation scheme of an industrial building.

In **stage 2**, the specification is expressed in the selected formal and semi-formal languages.

In **stage 3**, the specifications are translated, line-by-line, into the syntax of the Prolog reference language, assigning a legal Prolog name to each concept of each specification language. This is called the 'High-level Prolog' image.

In **stage 4** the Prolog specifications are expanded into the complete Prolog logic text, where the mathematical concepts implicit in the symbols are explicitly expressed. This is called the 'Full Prolog' image.

In **stage 5**, measures are obtained for both High-level and Full Prolog versions of the specification. The measures are taken from the text, directly observing the properties of the smallest meaningful Prolog entities, and composing from these the measures of all larger Prolog objects, using the 'algebra of measures' developed in Chapter 9 with an automated tool.

As is customary in presenting an experiment, Chapter 10 gives:

- the 'apparatus' used in the experiment (the selected specification languages, the translators, the Mathematical Toolkit for obtaining the Full Prolog from the High-level image, and the measurement tool PDQ for deriving the measures),
- the experimental method (the multiple-language strategy), and
- the summary of the results (the various versions of the specification as well as their measures).

The complete reference texts of the specifications, and their detailed measures, are given in the Appendix to Chapter 10, showing the build-up of the measures through the various phases of composition.

The interpretation of the measured results is also discussed in Chapter 10. The measures reflect several factors, some of which relate to the original referent, the specifier's modelling habits, the various specification languages, and the implementation of the underlying mathematical concepts. Each of these factors can be of importance according to context, and measures for them would be useful, but to appraise their effect it is necessary to isolate them. Chapter 10 suggests ways in which these factors might be isolated and their interaction reduced.

As an example of two of the factors of possible interest, consider first the measures of two sets of Prolog logic texts: High-level and Full Prolog. The difference between these two measures, for a given language and given specification, represents the difference between the level of the specification language and the level of the Prolog reference language. This level difference is a feature of the language, and can be used to characterize each specification language. On the other hand, if the object of interest is the original referent rather than the language in which it is expressed, then the measurement of its characteristics is best based on the Full Prolog image, which minimizes language dependence.

The models of Chapter 9 would allow further indirect measures to be derived, within the model-based measurement scheme, and utility measures to be composed over them. However, since our purpose is to demonstrate the construction of a measurement scheme, rather than to develop definitive measures for formal specifications, these developments fall outside the scope of the experiment, and of the book itself.

Note here that the models of Chapter 8 relate to the structure of Prolog, and are necessary in constructing Prolog's data measures in Chapter 9. However, the measures of Prolog structure, which characterize the *design* of the Prolog program, are irrelevant to the measurement of the specifications expressed in the Prolog text. The structural measures of Chapter 8 are included here for completeness, and further reference. These measures can be incorporated in some extended measurement scheme.

6.7 Summary and conclusions

A model-based measurement strategy has been proposed for formal and semi-formal declarative specifications, as a case study to demonstrate the feasibility and procedure of devising model-based measurement schemes for new fields. The primary aim is to illustrate the methodological value of model-based measurement, but the study has practical significance for the application domain itself. The exercise described here and detailed in Part 2 of the book, and the limited-scale validation experiments carried out in industrial contexts, indicate that a model-based measurement scheme can be readily developed, from first principles, for a large class of complicated referents, even if they were previously untouched by measurement.

In the application domain of declarative specifications, the demonstration study throws light on the compound modelling sequence involved in measurement: from original referent to specification, and thence to model-based measurement. In this application area measurement is further complicated by the diversity of the specification media which must be taken into consideration at this stage of development of the technology. The need for formal specifications is increasingly recognized and their use is demanded in some standards, but there are many formal and semi-formal specification languages on offer, a consensus about them is yet to arise, and it would be premature to tie the measurement strategy to a single formal or semi-formal language. In spite of such inherent difficulties of the application domain, even at this early stage of development of the model-based measurement scheme, some promising measures could be defined. In the study, the measures were not chosen arbitrarily, but were defined by articulating implicit characteristics in the model.

The case study demonstration, described here and in Part 2 of the book and the attended references, offers scope for research into the further development of model-based measurement of specifications, using it in quality assurance at the critical early phases of the life history process, such as characterizing *specifications* of complete systems and their high-level components. There is also scope for other important uses, such as characterizing and comparatively appraising *specification media* and defining *specification styles*. As part of such work, full-scale industrial validation of measures must be undertaken, and this would be facilitated by the availability of the experimental automated tools, developed for the demonstration exercise, and outlined in Part 2.

The demonstration study also shows the way ahead for devising model-based measurement schemes for other new application areas, and some such developments have recently been undertaken. Among these is the use of model-based measurement in project management, including the appraisal of project proposals and plans, and characterization of referents incorporating both hardware and software elements.

6.8 References

1   Goldschlager L, Lister A (1982): "Computer Science: A Modern Introduction." Prentice Hall International.

2   Tanenbaum A (1984): "Structured computer organization", Prentice Hall International.

3   Proc. 1st International Conference on Practical Application of Prolog – Section: "Prolog 1000", April 1992, London. Institute of Civil Engineers.

4 a Cunningham R, Finkelstein A, Goldsack S, Maibaum T, and Potts C (1986): Formal Requirements Specification – The FOREST Project. Proc IEEE Third InternationalWorkshop on Specification and Design, pp 186-192. IEEE Comp Soc Press.

b Knott R D, P J Krause (1988): "An approach to animating Z using Prolog". Report No A1.1, Alvey Project SE/065.

c White A P (1983): "Evaluation of Prolog as a rapid prototyping language for formal specification". MSc dissertation, Manchester University.

d Papapanagiotakis G, Azema P, Pradin-Chezalviel B (1986): "Propositional branching time temporal logic in Prolog". (Lab d'Autom. et d'Anal. des System. CNRS, Paris, France. 5th Annual International Phoenix Conference on Computers and Communication PCCCC'86 1986 Conf. Proceedings Scottsdale, AZ USA March 1986 Washington DC USA IEEE Comp Soc Press, p 371-7.

e Goble T (1989): "Structured systems analysis through Prolog". Prentice Hall.

5 Kowalski R (1974): "Predicate logic as a programming language". Proceedings of the IFIP Congress, Stockholm, Amsterdam, Holland, pp 569-574, North-Holland.

6 Robinson J A (1965): "A machine-oriented logic based on the resolution principle". Journal of the ACM, Vol. 12, No. 1, pp 23-41.

7 D H D Warren (1977): "Implementing Prolog – compiling logic programs". Research Reports 39 and 40, Dept. of AI, University of Edinburgh.

8 Roussel P (1975): Prolog: "Manuel de Reference et d'Utilisation. Groupe d'Intelligence Artificielle", Universite d'Aix-Marseille, Luminy, Sept. 1975.

9 Hogger C (1990): "Essentials of Logic Programming", Clarenden Press.

10a Bratko, I (1986): "Prolog Programming for Artificial Intelligence". Addison-Wesley.

b Clocksin W F, Mellish C S (1984): "Programming in Prolog". Springer-Verlag.

c Stirling L, Shapiro E (1986): "The art of Prolog". MIT Press, Cambridge, Mass.

11 van Emden M H, Kowalski R A (1976): "The semantics of predicate logic as programming language" Journal of the ACM 23(4), pp 733-742.

12 Hogger C (1987): "Prolog and Software Engineering". Microprocess and Microsystem, GB vol 11 No 6 p 308-18 July-Aug.

13 Kowalski, R (1979): "Logic for Problem Solving". North-Holland.

14 Draft expected.

PART 2

7 A FORMAL SPECIFICATION MEDIUM

7.1 Introduction

To demonstrate the multiple-language measurement strategy described in Chapter 6, we must choose and use a reference language. Prolog has been shown to be suitable for the purpose, and although it is not our aim to teach the reader Prolog, we must explain its properties in just sufficient detail to demonstrate its role in the measurement strategy.

Prolog is a formal language, capable of describing specification and design. Prolog's associated reasoning system can deduce the descriptive specification of the referent from its design, and confirm (or deny) its consistency with the prescribed specification. In practice, Prolog is used not only for this purpose but also as a programming language, hence its facilities extend beyond those required for specification and design.

This chapter first presents Prolog as a programming language, examining the different components of Prolog and their role in the language. It goes on to show how the logic component, that part of the Prolog text relevant for expressing specifications and designs, may be abstracted. The non-logical components which are removed have measurable features of their own, and these would need to be incorporated into some future measurement scheme for characterizing Prolog as a programming language.

Section 7.2 shows how Prolog is related to logic, and illustrates some simple Prolog specifications in order to acquaint the reader with the syntax of Prolog through examples, before proceeding to the modelling of the text. Section 7.3 subjects the Prolog program text to a series of abstractions, isolating the logic text. In Section 7.4 this Prolog logic text is treated as a referent whose properties are to be captured by model-based measurement. Two aspects are considered: the hierarchical structure of Prolog's logic programs, irrespective of the data which flows through the structure, and the properties of the data itself. The model-based measurement scheme developed in Chapters 8 and 9 concentrates on these two aspects. Section 7.5 expands on the notion of levels of language mentioned in Chapter 6. Section 7.6 contains a summary.

7.2 Logic, Prolog and some specifications

7.2.1 Formal logic into Prolog

Formal specifications may be expressed in the following format: a specification is named, and the name is followed by a list of conditions which must be satisfied for the specification to hold. In general, the conditions are themselves expressed in terms of other attributes, and the specification can be given in the form:

condition$_1$(argument list$_1$) AND condition$_2$(argument list$_2$),
$$\text{AND} \ldots \text{condition}_n(\text{argument list}_n).$$

In this expression each condition is a *predicate*: a Boolean function whose value is *true* or *false*, as determined by the values of the variables in the argument list. Using the notation of predicate logic ($\wedge$ for AND and $\Leftarrow$ for implication), and abbreviating 'argument list' as 'arglist', we write:

specification(arglist) $\Leftarrow$ condition$_1$(arglist$_1$) $\wedge$ condition$_2$(arglist$_1$)
$$\wedge \ldots \text{condition}_n(\text{arglist}_n)$$

We may now convert this to Prolog syntax, retaining only the shorthand form Arglist:

specification(Arglist) :- condition$_1$(Arglist$_1$), condition$_2$(Arglist$_2$), ...,
$$\text{condition}_n(\text{Arglist}_n).$$

This Prolog clause reads as 'specification is true if condition$_1$ is true and condition$_2$ up to condition$_n$ are true'. Each of the conditions (or antecedents), appearing on the right hand side of the implication, is a sub-specification. Each of these is a specification in its own right. It may be the specification of a component of the system, or the specification of a mathematical tool, or a Prolog system predicate. It is usual to call the consequent, on the left of the implication, the head of the clause, and the list of antecedents the body. In Prolog there is always only one predicate on the left of the implication.

7.2.2 Examples of Prolog specifications

Three simple examples are presented to introduce Prolog as a medium of specification, and illustrate the reading of Prolog specifications, both declaratively and procedurally. The examples are: list membership, factorial, and the ancestor predicate . (Note that in the Prolog literature the terms 'predicate' and 'relation' are used synonymously. We use 'predicate' in the context of Prolog, and reserve 'relation' for describing compositions over elements of sets.)

EXAMPLE 'member/2'

Membership of a list may be specified by the following:

```
member(X, [X|Tail]).                          % This is a comment
member(X, [Y|Tail]):- member(X, Tail).
```

This portion of Prolog code has two clauses, each terminating in a period. The first clause is a *fact clause*. It consists of one predicate, of arity 2. 'member' is the name symbol of the predicate. The two terms consist of variables, denoted by a sequence of alphanumeric symbols whose first character must be upper-case alphabetic. The second term has two variables enclosed in the 'list' symbols. The first variable, X, signifies the first element of the list. The second variable 'Tail' is itself a list: the remainder, excluding X.

The second clause is a *rule clause*. Its Head is a single predicate and its Body is also a single predicate. This second clause is recursive since its Body contains a predicate with the same name symbol and arity as the Head. Together, the two clauses form a procedure, since their Heads have the same name symbol and the same arity. Adding the 'start' predicate, ?member(A,B), with either A or B or both instantiated, converts the text to a program.

Reading **declaratively**, the first clause states that X is a member of a list if it matches the head of the list. Variable 'matching' is implied by the use of the same variable symbol X. This is a fact clause with no conditions. Provided the same symbol can represent the element and the head of the list, the clause is always true. The second clause states that X is a member of a list if it is not the first element in the list and yet is a member of the remainder of the list (Tail). The procedure 'member/2' comprises the two clauses of that name, in disjunctive (OR) relation. In the query ?member(X,List), X is a member of List if either of the clauses evaluates to 'true'.

The case of X=Y is filtered out by the first clause. In cases where ordering cannot be taken for granted, as in concurrent Prolog, a guard would be added to the second clause, thus:

```
member(X, [Y|Tail]):- X \== Y, member(X, Tail).
```

In this case the second body predicate, 'member(X,Tail)' is guarded by the first filtering out the case that X is not identical to Y. With the clause in this form, the order of clauses becomes irrelevant.

The **procedural** reading of the same procedure is: to test if X is a member of a list, first see if it is the first element of the list and if not, remove that element and test the remainder.

END OF EXAMPLE 'MEMBER/2'

EXAMPLE 'factorial/2'

The factorial of a positive integer n (the product of all positive integers up to and including n) is defined in Prolog as:

```
factorial(0,1).
factorial(X,Y):- X1 is X - 1, factorial(X1,Y1), Y is Y1*X.
```

The first clause, a fact clause, states that factorial 0 is 1– a definition.

The second clause states that factorial X is Y if factorial X-1 is Y1 and Y is the product of Y1 and X (where the infix predicate 'is/2' forces evaluation of X1). In this version X must be a non-negative integer.

END OF EXAMPLE 'FACTORIAL/2'

EXAMPLE 'ancestor/2'

A definition of the ancestor predicate is

```
ancestor(X,Z):- parent(X,Z).
ancestor(X,Z):- parent(X,Y),ancestor(Y,Z).
```

The first of the two clauses states that X is an ancestor of Z if X is a parent of Z. The second clause states that X is an ancestor of Z if X is a parent of Y and Y is an ancestor of Z. Given a series of fact clauses detailing the parent/child predicates, this will eventually work down the family tree of X until it reaches Z or fails. Failure in Prolog is negation. There is no 'not proven' verdict. If the query evokes the response 'no' then, *within the knowledge base of the program*, X is not an ancestor of Z.

END OF EXAMPLE 'ANCESTOR/2'

These examples show how Prolog may be used for specification. The fact that each of the examples will also produce a solution to a query is, from a pure specification outlook, a bonus. If the order of the clauses in example 'factorial' is reversed, no solution will be forthcoming, as the program will not terminate. The compiler will keep decrementing X by one, and trying again, being constrained not to try a second clause until all possibilities on the first are exhausted. This will not happen unless negative values of X are prohibited, by the addition of some extra code. In example 'ancestor', too, if the order of the body predicates is reversed, the program will not terminate. However, all three are satisfactory declarative specifications of their respective problems, whichever way their clauses are ordered. Ordering of clauses represents one aspect of the conversion of a declarative specification to an executable program.

7.3 Components of the Prolog logic text

The essential part of the syntax of a programming language, which allows the formulation of meaningful statements, is sometimes referred to as the *abstract syntax* ([1]). In addition to this, the syntax of the language may make provision for utilities which extend the syntax into an efficient operational programming language. In ([1]),this is referred to as the *concrete syntax* of the language. It is usual also to provide further facilities for comments, and the commented text is sometimes called the *decorated program text*.

In the case of Prolog, the abstract syntax is identified as the *logic* component of the declarative program text. This is the essential part of the program for expressing the facts and rules which specify the problem and link its specification to its design. Additional elements of program text are needed to augment the abstract syntax to the full syntax of the procedural Prolog programming language. These extra elements provide a set of *utilities*. Prolog's utilities may conveniently divide into two groups: control utilities and system predicates. These overlap somewhat, since some control utilities are implemented by system predicates. The full Prolog program text can be further supplemented by *comments* which are, strictly speaking, outside the syntax of the language, but facilitate its use.

This section acquaints the reader with Prolog, presents the various components of the complete 'decorated' Prolog text, and discusses their function. With the aid of Figure 7.1, and in the order shown in the figure, it explains how the decorated Prolog program text may be stripped away systematically, stage-by-stage. What remains is the logical part of the declarative Prolog text – the part needed for expressing specifications and designs. This 'Prolog logic text', which is our object for model-based measurement, may or may not be executable.

7.3.1 Comments

Comments are informal statements, outside the syntax of the programming language, ignored by the language translator (the compiler or interpreter). Comments are inserted into the text at the time of writing the program, in order to facilitate reading. They are reminders for the program's author, and explain his/her intentions to other readers. The comment-free program may be correct, but its meaning may be obscure. Bad commenting merely repeats what is already obvious from the code. Good commenting offers a higher-level picture: it provides a wider perspective, and gives pointers to other items in the code. The main use of comments is to aid quality assurance, and facilitate the modification and maintenance of software. Commenting is linked with such software properties as understandability and readability.

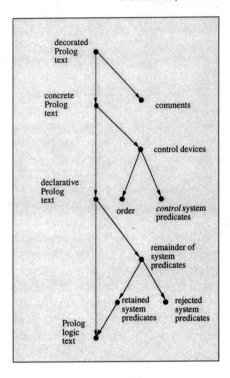

Figure 7.1: Preliminary decomposition of Prolog

After removing comments, the remainder is the 'concrete Prolog text', shown in Figure 7.1. Neither the logic of the declarative program, nor the executibility of the procedural code is affected. From the viewpoint of measurement of specifications and designs, comments are not of immediate interest. Should models and measures of the comments themselves be necessary, the abstracted comments may be linked to the structural model developed in Chapter 8. Suggestions for good commenting are given in ([2]).

7.3.2 Control utilities

Control utilities of Prolog programs guide the interpreter/compiler as to the order in which procedures and clauses should be processed in course of searching for logical resolution of the conditions specified in the program. (For resolution, see [3], [4]). The guidance to the language translator will be for the purposes of avoiding non-terminating searches, and increasing efficiency.

The main control element in Prolog is implemented by the ordering of the clauses in a procedure, and of the body predicates within the antecedents of each clause. Since logic is commutative, the order of statements is no part of the abstract syntax of the declarative Prolog program. To isolate this element of the concrete syntax, we simply declare that the sequence in which statements occur in the program text is arbitrary. For model-based measurement, this abstraction carries the obligation that the model must preserve meaning, irrespective of the order of the statements of the text, and the measures must also be invariant with order. In all subsequent examples of code and its graphical representation, we are at liberty to change the textual order of clauses within a procedure, and the textual order of body predicates within a clause.

Other means of control available to the Prolog programmer are four of the system predicates of Prolog: 'true', 'fail', 'repeat' and '!'. (The last of these is referred to as the 'cut'). Their function is operational: they influence search during the execution of the program, but contribute neither to the structure nor to the data. 'True' and 'fail' govern whether the search is to continue or backtrack, and 'repeat' acts as a barrier to backtracking. 'Cut' prunes the search tree. Its use is controversial, and it is often possible to avoid it ([5]). The four controlling system predicates are easily abstracted from the code, leaving behind the declarative Prolog text (Figure 7.1). Removing these system predicates will almost certainly affect run-time characteristics for ill.

Measures for the frequency and distribution of the control predicates could be tied to the structural measures of Chapter 8, as in the case of comments. Should measures of operational efficiency be required, this would call for modelling the dynamic search tree, and the effects which order and control predicates have upon it.

7.3.3 System Predicates

System predicates – other than the controlling system predicates already discussed – have no overt definitions in the text, and, provided they are syntactically correct, they evaluate to 'true'. They achieve their purpose through 'side effects'. For example, the predicate 'write('Enter password')' will evaluate to 'true', and, as a side effect, will put up the appropriate message on the screen. System predicates fall into broad groups (see e.g. [6]), but there is no consensus on the criteria of classification and the exact composition of these groups. Their number and form will depend on the dialect of Prolog.

Most system predicates are not concerned with the logic of the program but handle such practical matters as peripheral interchange (reading in programs, file handling and input/output), arithmetic, debugging, and modification of the program via the utilities 'assert/1' and 'retract/1'. Most of these will be removed when abstracting the logic component of the Prolog (Figure 7.1).

A small number of system predicates are considered essential to the logic of the program because the concepts they represent, those of comparison and set theory, are basic to specification of systems. These are of two sorts: comparison operators, and some of those predicates which contain 'metavariables' in their arguments, the metalogical system predicates. The comparison operators are: <, >, =, ==, and their various combinations. They are all binary. The retained metalogical predicates are: 'setof/3', 'bagof/3' and 'forall/2'.

The inclusion of this selection of system predicates is an example of the modeller imposing a subjective decision on the model-based measurement scheme. The modelling decision is always available for reassessment in the light of experience. Should other system predicates prove necessary during the course of the work on translation of specification, these can easily be included in an extended version of the current model-based measurement scheme.

7.3.4 Prolog logic text and its role in model-based measurement

After abstracting away the three language components (comments, control and selected system predicates), what remains is the logic part of the Prolog text (Figure 7.1). This comprises a set of 'facts' and the 'rules' linking them, together with those system predicates identified above which are deemed to belong to the logic text.

As we shall see in the next section, the Prolog logic text may be a program, capable of representing the specification and design of the original referent, or only part of a program, but still a coherent language entity. This is the *referent*, to be subjected to model-based measurement, in accord with the procedure discussed in Part 1, and reproduced for the present case in the accompanying Figures 7.2. and 7.3. (Validation loops are omitted.)

Note now the composite process of model-based measurement at work. Inside the process, so to speak, our referent is an arbitrary segment of Prolog logic text, as shown in Figure 7.2. Let us assume that this is a program. But then the Prolog logic text is itself a model: the representation of some real-life referent under specification and design. Within the measurement scheme, modelling is a two-step process: first, the Prolog logic text is created as a model of the true, primary, real-life referent, and then the logic text is treated as a secondary referent, to be modelled and measured (see Figure 7.3). The measures characterize the secondary referent – the Prolog logic text – and through this they are interpreted as the attributes of the primary referent – the entity under specification and design.

In the rest of this chapter, and in the subsequent two chapters, we treat the Prolog logic text as the referent of the model-based measurement scheme, in accord with Figure 7.2. We then return to the wider view, given in Figure 7.3, examining the way in which the measures of the Prolog logic text reflect the attributes of the real-life referent itself.

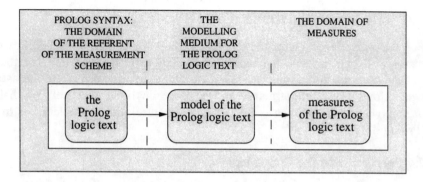

Figure 7.2: Outline of model-based measurement of the Prolog logic text

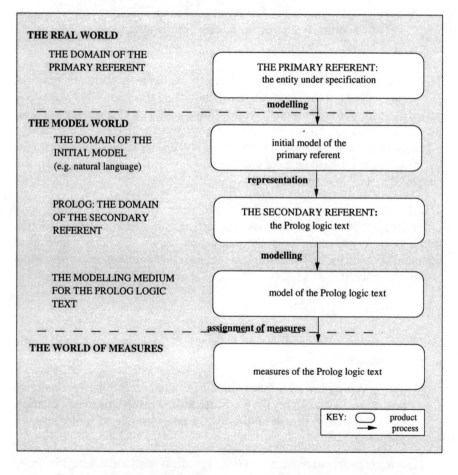

Figure 7.3: Model-based measurement of a real-life referent through measurement of its representation in Prolog logic text

7.4 The referent: the Prolog logic text

For the purpose of characterizing the Prolog logic text, it is helpful to distinguish two aspects: the logical interrelationships which link together Prolog relations into a multi-level logical hierarchy, irrespective of the data which flows through the structure, and the properties of the data itself. We shall be referring to the former as the 'Prolog structure' and to the latter as the 'Prolog data'.

7.4.1 Structure in the Prolog logic text

We model Prolog structures in the usual way:

$$M_S(Prolog_s) = (\mathbf{Comp}_s, \mathbf{R}_{Comps}) \dots\dots\dots\dots\dots\dots\dots\dots Eq\ 7.1$$

where   (Prolog$_S$) is the Prolog structure,
             Comp$_S$ is the set of elements of the structure

and   **R**$_{Comps}$ is the set of relations holding over the element set.

The practical way of obtaining a model of structure is to retain only the predicate symbols of the logic program, discarding the data. What remains is called the 'stripped code' model ([7]) of the logic program. This section provides definitions of these components and of the relations which knit them into the logic component of a complete Prolog text. It is this structure which forms the basis of the model-based measurement scheme to be developed in subsequent chapters.

The definitions are given 'top down' and recursively. Where terminology differs from that of standard books on Prolog, this is due to the restriction imposed by considering only the static text, which excludes non-logical components such as 'call'. All implementation considerations and operational issues are ignored.

Types of Prolog logic text

We distinguish two sorts of Prolog logic texts: 'logic programs' and 'logic program segments'.

(1) A *logic program* is a (start) predicate and a matching hierarchy of procedures which imply the predicate and are resolvable to fact-clauses. (It is implicit in this definition that a program is also a hierarchy of programs.)

(2) A *logic program segment* is a (start) predicate and a matching hierarchy of procedures which imply the predicate such that some or all procedures are not resolvable to fact-clauses.

Note on resolution

(i) Procedures P_1 and P_2 form a two-level *hierarchy* if the head of P_2 matches one of the body predicates of P_1. P_1 is then the higher level procedure.

(ii) A procedure is *resolvable* to fact-clauses if each of its clauses

    a) is disjunctively composed of fact-clauses, or

    b) is hierarchically composed of procedures which are themselves resolvable to fact-clauses.

A rule-clause is resolvable to fact-clauses if each predicate of its body has a procedure head where a procedure head is a disjunction of matching clause heads, each resolvable to fact-clauses.

(iii) Predicates *match* if they have the same name and the same arity.

Clauses match if their head predicates match.

A procedure matches a predicate on the body of a clause if its head matches the body predicate.

End of Note on resolution

Components of the Prolog logic text

The Prolog logic text is composed of three kinds of parts. These are members of the component set **Comp$_s$** of the Prolog logic text.

(1) A *procedure* is a disjunction of matching clauses.

(2) A *clause* is composed of a *head predicate* and a sequence of *body predicates*, linked by implication from body to head. There are two kinds of clauses, fact-clauses and rule-clauses.

- A *fact-clause* is unconditionally true, hence its body is empty:
- A *rule-clause* is a clause whose head is a predicate and whose body is composed as a conjunction (AND function) of predicates.

(3) A *predicate* is a name symbol, followed by the data associated with the predicate. From the viewpoint of resolution there are two types of predicate:

- those explicitly resolvable predicates which have a matching clause in the program,
- and a further type which are implicitly resolvable 'system predicates' having a concealed matching clause within the Prolog compiler/interpreter.

Relations of the Prolog logic text

The four structural relations which make up the relation set **R**Comps of the Prolog logic text are conjunction, clausal implication, disjunction and hierarchical implication.

(1) *Conjunction* (AND relation) is applied over predicates to form a clause body.
(2) *Clausal implication* acts from body to head, linking the two.
(3) *Disjunction* (OR relation) is applied over matching clauses to form a procedure.
(4) *Hierarchical implication*, from a procedure to a predicate matching the procedure head in a clause body, links together adjacent levels of the hierarchy.

Figures 7.4a illustrates the simplest possible, single level, logic program. (The start predicate is not considered to be a countable level). In the logic text, the start symbol 'Predicate' is implied by a procedure consisting of a series of fact clauses. This is modelled by an arrow whose meaning is shown in italics. The fact clauses are linked by *disjunction*. The program is thus resolvable to fact clauses. Programs may have a *conjunction* of start predicates, when the program would be represented by more than one such tree. (For a graphical representations see Figure 7.5c, considering only levels 0 and 1. In that figure the arrows are reversed and model the 'implies' relation rather than is_implied_by.)

Figure 7.4b illustrates the structure of the top level of a multi-level textual program. The figures show, in italics, the meaning of the arrows, and, in capitals, the composition relations, in addition to the usual Prolog symbols. The procedure implied by the start symbol now contains, in place of just fact clauses, a disjunction of 0 or more fact clauses and 1 or more rule clauses. With no rule clause and one or more fact clauses, the figure degenerates to that of Figure 7.4a. Each rule clause contains within its body, a *conjunction* of predicates, each of these being a complete program in itself – labelled sub-program in the figure. For simplicity the figure expands only one of the rule clauses. It is important to note that, although the head of a clause appears in the figure and is used in the execution of programs, from the point of view of textual Prolog it does not form part of the hierarchy of the figure.

The tree structure of Figure 7.4b conceals the fact that sub-programs and procedures may appear at more than one level, being their own ancestors. This introduces cycles, and the model becomes a cyclic digraph. This is called recursion. Any modelling process based on a hierarchical tree structure must take into account the cycles created by recursion. Recursion is illustrated in Figure 7.4c.

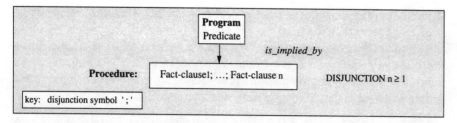

Figure 7. 4a: Simplest textual Prolog logic program, resolved to fact-clauses

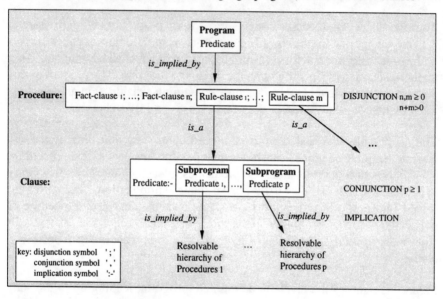

Fig 7.4b: Hierarchical structure of Prolog logic program

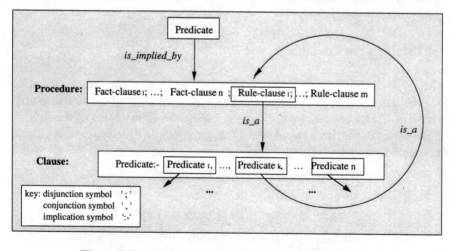

Figure 7.4c: Simple recursion in Prolog logic program structure

A recursive clause of special interest here is one which contains in its body no predicates other than the head predicate, e.g. of the type 'pred:-pred.' In order to terminate, the procedure must also contain a fact clause 'pred.' representing the boundary value. Such procedures are resolvable to their boundary fact clauses, and, depending on no other procedures, are self-contained programs in their own right (see Example program 'member/2').

Building a logic hierarchy

The level-by-level process of composing a Prolog program as a logic hierarchy is enlarged upon in the series of Figures 7.5a - 7.5c. The figures show the role of the four basic constructors in building the program from the bottom up, from its fact-clauses, and finishing with the complete program. The product of this process is a multi-level nested structure, containing a disjunction of pathways. Each disjunction may contain both fact-clauses and rule-clauses.

The arrows are reversed for this 'bottom-up' process and now model the relation 'implies'. Figure 7.5a shows the operators inside a single procedure. The conjunction operator links the predicates in each clause body. The body of each clause is then linked to the head by clausal implication. The clauses are joined into a single procedure by disjunction and the whole procedure is connected to lower and higher levels by hierarchical implication. Figure 7.5b shows the special hierarchical link in a procedure containing a recursive clause.

Figure 7.5c shows the levels in a complete program, leading up to the start symbol. By definition, the lowest level can have no in-arrows from below. This level consists of procedures of fact clauses and the self-contained recursive clauses described above.

7.4.2 Data in the Prolog logic text

The scope of a Prolog variable is confined to one clause. The clausal structure of Prolog is described in Section 7.3.1 as a structure of predicates, containing a head predicate and a sequence of body predicates. The totality of the data of these predicates is the 'clausal data'. The data domain of a clause in Prolog is modelled as a system:

$$M_S(Prolog_D) = (Comp_D, R_{Comp_D}) \quad\quad\quad\quad\quad Eq. 7.2$$

where $Prolog_D$ is the Prolog clausal data,
            $Comp_D$ is the set of data elements in the clausal data,

and R_{Comp_D} is the set of relations holding over members of the clausal data set.

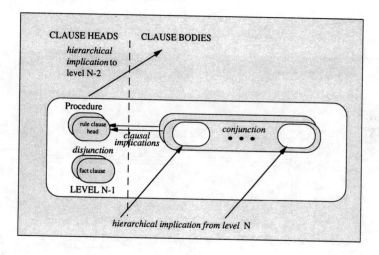

Figure 7.5a: Composition operators for procedures, excluding recursive clauses

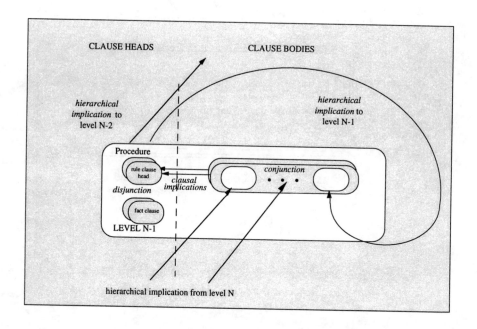

Figure 7.5b: Composition operators for procedure, including a recursive clause

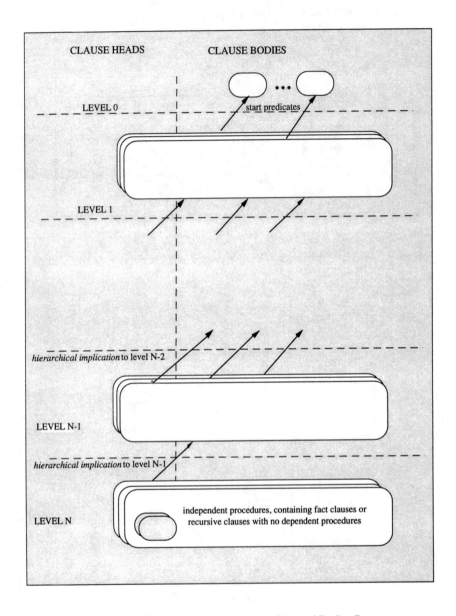

Figure 7.5c: Hierarchical composition of textual Prolog Program

Components of the clausal data

The *clausal data* is composed of a set of *predicate data*, where the set comprises all the predicates of the clause.

Predicate data is a sequence of *data objects*.

A *data object* is composed of an argument and a term.

An *argument* is a location for a term.

A *term* is either compound or atomic.

- A *compound term* (also called a functor) comprises a name and a sequence of data objects, ultimately composed of atomic terms.

- An *atomic term* can be a *constant* or a *variable*.

    * *Constants* are denoted by lower case letters and numerals, e.g. 'smith', 'jones', '42'.

    * *Variables* are denoted by initial upper case letters, e.g. 'X', 'Variable', etc. Strictly speaking, Prolog variables should be called 'identifiers' since, within their scope of a clause, and after being bound to a value, they cannot be varied. Use of the word 'variable' is widespread, and we conform to this usage. There is one special variable, the understroke or '_'. This is called the *don't care* variable. It is never bound to a value.

Note on functors

In Prolog, a functor is distinguished from a predicate only by its context, which is as a compound term within a predicate argument. Take for example the construct:

    address(Street1,Street2,Town,PostCode)

This is a syntactically correct Prolog predicate. However, when used as the contents of a predicate argument such as:

    find(Name, address(Street1,Street2,Town, PostCode)) :- <suitable body>,

it is recognized as a functor, a compound term.

A special type of functor is the binary list functor '.(Head,Tail)'. Read this as 'dot-Head-Tail'. The functor is special in that Tail can only be a list. Thus, introducing further elements into a list leads to multiple nestings. This notation proved cumbersome and hence has been replaced by the format: [Head|Tail]. In this notation, adding elements to the list does not involve nesting. Contrast .(a, .(b, .(c,Tail))) with [a, b,c|Tail]. Only the second notation will be used in the remainder of the book. User-defined functor symbols are lower case. A functor with no arguments thus degenerates to a Prolog constant. Hence an empty list, [], having no arguments, is a constant. Another special type of functor is allowed in some Prolog dialects. This is the n-tuple, (A,B, ..., N), with no explicit name.

End of note on functors

Relations of the clausal data domain

The non-atomic data objects of Prolog are composed by the following data relations which make up the set $\mathbf{R}_{Comp}$.

The *identity* relation holds over the set of variables of the clause which bear the same name and binds them to the same value.

The *sequence* over data objects of a predicate is an ordering relation.

The *contains* relation links an argument to a term.

The *constructor* relation creates a compound term as a sequence of data objects linked to a name.

Linking data objects of different clausal domains

Inherited from program structure are the relations of disjunction and hierarchical implication which provide the means of composing the data of clauses of a complete program across the clausal boundaries.

Hierarchical implication allows transmission of information between the data of two clauses: entities of distinct scope. The data of the clause head on some level M and the clause body on the higher level M-1 are merged. Given a match between clause head and body predicate, information channels are formed through position matching on the arguments. (Prolog users will recognise this as *unification* [4]. The process of unification need not be considered in the context of static code.) The program segment below provides an example in which A is matched with F, and L is matched with [X,Y].

FAMILY EXAMPLE

This simple example speaks for itself, demonstrating how hierarchical implication operates, forming the link across logic levels.

```
mother(M,L):- spouse(M, A),
              children_of( A, L).



children_of(F, [X,Y]) :-
            father(F, X),
            father(F, Y).
```

END OF FAMILY EXAMPLE

7.4.3 Modelling the referent

Figure 7.1 showed how the Prolog logic text is isolated in preparation for model-based measurement. The process continued, creating two types of models of the Prolog logic text, to allow each of them to be modelled and measured independently. Figure 7.6 continues from the final node of the earlier Figure 7.1 to show two possible abstractions of the Prolog logic text.

- The stripped code model is based on abstracting only the predicate symbols and discarding the data. This model is developed and measured in Chapter 8.
- The second abstraction, that of the *data*, discards the predicate symbol names and retains the data. If the Prolog logic text is a single clause then this is the only part required. If the Prolog logic text is more than a single clause then the data model draws on the structural model in order to cross the clausal boundaries. The model and measures of the data are presented in Chapter 9.

In the case of Prolog as a specification medium and as a unifying reference language, we start with Prolog logic text, hence Figure 7.6 is sufficient to represent preliminaries to the modelling and measurement method.

7.5 Language levels of formal specifications

As described in Chapter 6, most specification languages operate at a higher (less detailed) level than Prolog. Translation into Prolog as the unifying reference language forces us to identify an *implicit* distinction between:

- high level notions into which these higher-level specification languages are directly translatable;
- low-level notions which articulate the meaning of the mathematical symbols used in the high-level specification language.

Figure 7.7 shows graphically the level of a specification language other than Prolog, and its equivalent in Prolog. A line-by-line translation of a high level language into Prolog will produce text at an equivalent level, lacking the substructure that would make it complete in itself. For example, should a specification contain a line 'A $\subseteq$ B' then the Prolog equivalent will be subset(A,B). This statement on its own is named a High level Prolog statement. It assumes that the concept implicit in the subset symbol of the higher level specification language is also implicit in subset(_,_). However, from a Prolog point of view, subset(A,B) is meaningless unless a procedure which defines 'subset/2' is included. This is consistent with the fact that Prolog is a low level

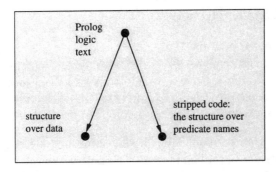

Figure 7.6: Two abstractions on the Prolog logic text

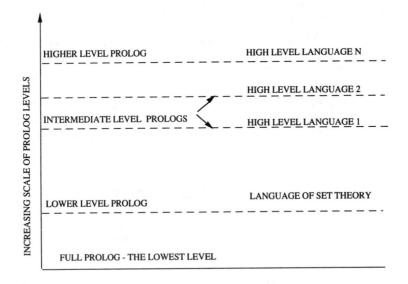

Figure 7.7: Levels of Prolog

language requiring more detail than a higher level language. Figure 7.7 shows how each specification language relates to its own level of Prolog. The richer the language symbol set, the higher will be the level of its straight translation into Prolog.

Based on the distinction between levels, the translation of formal specifications into Prolog is thus seen as a two-stage process. The first stage merely replaces the mathematical symbols of the original specification by a Prolog equivalent. To complete a specification, in the second stage both the mathematical symbols of the specification and the Prolog equivalent would need to be amplified, the former by a glossary or a standard mathematical textbook, the latter by a defined procedure or a system predicate.

In the translation of formal specifications directly into High Level Prolog, since no detail is added to the translation and no detail removed, the two texts, that of the specification language and that of the comparable Prolog, are isomorphic.

We call that version of the Prolog text which is sufficiently detailed to be acceptable to a compiler or interpreter 'Full Prolog'. The allowed system predicates of Full Prolog need to be tied to a particular version since these vary slightly amongst themselves. We await the ISO standard, and meanwhile accept the Edinburgh syntax as standard.

Producing a Full Prolog version of a specification from its High level version amounts to adding specifications for the mathematical symbols used. We can simplify the process by producing one large collection of symbol specifications to cover all the symbols used in the various specification languages of interest. Specifications in particular languages will then draw their own symbols from this super-set. Chapter 10 gives further details.

7.6 Summary

Prolog finds use as a programming language and as a declarative language for expressing and reasoning about specifications and designs. To use Prolog as the referent language in the multiple language measurement strategy, its logic component must be isolated. This is called the 'Prolog logic text'.

The specification/design under investigation must be described in the Prolog logic text. This representation is treated as a secondary referent, and is subjected to model-based measurement. If the models and measures are sound, these measures will be representative of the attributes of the original referent.

Two types of model of the Prolog logic text are developed: (1) the model of 'Prolog structure', which describes the logic text as a multilevel hierarchy resolved into predicates, and (2) the model of 'Prolog data', which represents the data items and their interrelationships. Since the scope of data items in Prolog are confined to a clause, the model of the Prolog structure is required for assembling the data model of a program, or any language entity larger than a clause.

With the aid of its hierarchical structure, Prolog admits the description of the original referent on an arbitrary level of detail. 'Full Prolog' – the lowest level version – contains the most detail. This versatility of Prolog is a valuable feature when implementing the multiple language strategy. It facilitates translation from any of the formal languages of the strategy into Prolog. Translation can proceed mechanistically, line-by-line, into a matching level of Prolog. For the sake of comparisons, the translation can then be expanded into Full Prolog.

7.7 References

1 McGettrick A D (1980): "The definition of programming languages". Cambridge University Press.

2 Benwood, H J J (1988): "A software tool for the structural measurement of Prolog". MSc dissertation, Heriot-Watt University.

3 Robinson J A (1965): "A machine-oriented logic based on the resolution principle", Journal of the ACM Vol 12, No. 1, pp 23-41.

4 Hogger C (1990): "Essentials of Logic Programming". Clarendon Press.

5 Van Le P (1993): "Techniques of Prolog Programming". Wiley.

6 Byrd L, Pereira, F C N, Pereira L M, Warren D H D (1982): "DECsystem-10 Prolog Users Manual". Department of Artificial Intelligence, University of Edinburgh.

7 Myers M (1986): MSc dissertation "The introduction of axiomatic theory of structures to Prolog". South Bank University.

8 MODELS AND MEASURES OF STRUCTURE

8.1 Introduction

The aim of this chapter is to present a model-based measurement scheme for characterizing the *structure* of Prolog logic texts. This chapter takes no account of the data content of the Prolog logic text, but concentrates only on the hierarchical relations in the structure, illustrated in figures 7.4a to 7.4c of Chapter 7.

In Sections 8.2 and 8.3 we develop the structural model, and show how to represent the structure of each Prolog construct, from the simple predicate through to any valid portion of logic text. We build the structural model of each construct, starting from the simplest – the predicate – through a sequence of predicates, a clause and a single procedure, to a hierarchy of procedures, including complete programs. Section 8.4 proposes structural measures for each of these, and 8.5 outlines a tool for automating the measurement.

8.2 Modelling the structure of Prolog logic text

The Prolog logic text is first abstracted from the full text, following the sequence of actions detailed in Figure 7.1. A further abstraction removes the data with its enclosing brackets. What remains is called the '*stripped text*' or '*stripped code*' ([1]). The stripped text is a model of the original Prolog, itself the referent for the further, graphical model developed in this section.

The stripping and labelling of program text is shown in Figure 8.1 on the EXAMPLE PROGRAM 'quicksort' ([2]). The stripped code is labelled for convenience. Each body predicate in a clause bears a number uniquely attached to that name. This number is also attached to the same predicate name when it is the head of a clause, but here, in addition, the number carries an alphabetical label, a, b, c, etc., according to which clause of the procedure it heads. The labels give a convenient shorthand for tracing back to predicate

```
:-quicksort(List,Sortedlist).          1.  :-quicksort.
quicksort([],[]).                      1a. quicksort.
quicksort([H|Tail],S):-                1b. quicksort:-
    split (H,Tail,A,B),                    2. split ,
    quicksort(A,A1),                       1. quicksort,
    quicksort(B,B1),                       1. quicksort,
    append(A1,[H|B1],S).                   3. append.

split(H,  [A|X],[A|Y],  Z):-           2a. split:-
    A  H,                                        ,
    split(H,X,Y,Z).                            2. split.
split(H,  [A|X],Y,  [A|Z]):-           2b. split:-
    A>H,                                       >,
    split(H,X,Y,Z).                            2. split.
split(_,  [], [], []).                 2c. split.

append([],L,L).                        3a. append.
append([X|L1],L2,[X|L3]):-             3b. append:-
    append(L1,L2,L3).                          3. append.
```

(a) Text for program 'quicksort'   (b) Labelled stripped text for
                                          program 'quicksort'

Figure 8.1: EXAMPLE PROGRAM 'quicksort' and its stripped and labelled text.

names. In Chapter 7 we discussed the elimination or retention of system predicates, coming to the conclusion that a small number, the comparison operators, should be retained in the data model. For consistency, this section retains those same system predicates.

8.2.1 Modelling the structural entities

The structural entities of Prolog text are first modelled as labelled digraphs. A further modelling step yields two abstractions: an *unlabelled digraph* and a *bag of labels*. Both serve as sources for the measures described in Section 8.4. In the examples that follow, generalized models are unlabelled. Instances of models are either labelled, or, if unlabelled, are accompanied by a bag of labels. There are two nodal symbols and an arc symbol used in the digraph, as shown in Figure 8.2.

Modelling predicates

Predicate names are first modelled as head-nodes or body-nodes on a digraph. The form of the node varies as the predicate name occurs in the head or body of a clause.

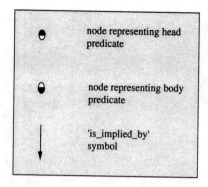

Figure 8.2: Symbols used in the digraphs

Modelling the start node

The start node is a body predicate and is modelled as a body-node.

Modelling clauses

A *fact clause* is modelled as a single head node. It is a feature of this model that a rule clause whose body predicates are all the 'rejected' system predicates, discussed in Chapter 7, is modelled as a fact clause.

A *rule clause* is modelled as a head node and a set of body nodes, joined by arcs from the head node to each of the body nodes. In the simple rule clause model, each path is of unit length, as in Figure 8.3a.

Figure 8.3b shows labelled models of the three clauses of procedure 'split' and the two clauses of procedure 'quicksort' in the example program listed in Figure 8.1. The nodes are identified by the labelling system used in producing the stripped text. The third clause of 'split' and the first clause of 'quicksort' are fact-clauses; the other clauses are rule clauses. Each of the rule clauses of 'split' has a retained body system predicate which is an infix comparison operator. These keep their original names. The unlabelled models, together with the bags of labels, are given in Figure 8.3c.

Figure 8.4 summarizes the characteristics of the clausal bipartite tree , together with the constraints imposed by the nature of the referent.

Modelling procedures

An isolated procedure is a disjunction of clauses. A procedure model is a forest, consisting of trees of clause models, Figure 8.5. The modelling process

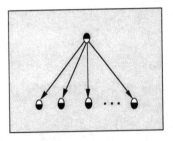

Figure 8.3a: Generalised clause model

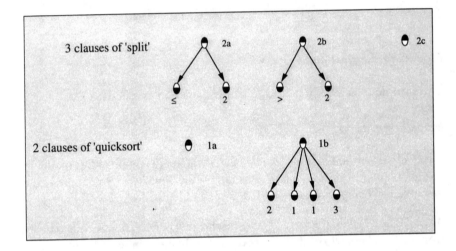

Figure 8.3b: Labelled models of fact- and rule-clauses

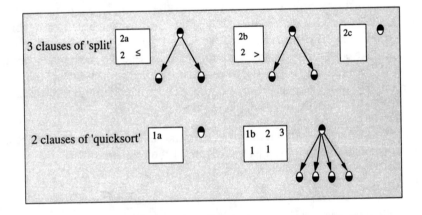

Figure 8.3c: Unlabelled clause models

feature	symbol	properties
predicate in head of clause	◑	zero in-arcs, zero out-arcs in fact clause, one or more out-arcs in rule-clause, labelled as alphabetically suffixed integers.
predicate in body of clause	◑	one in-arc, zero out-arcs, labelled as integers only.
'is-implied-by' relationship	↓	clausal implication, the leaves of the tree are implicitly joined by conjunction.
fact-clause	◑	single head predicate
rule-clause	see figure 8.3a	2-level bipartite tree, root is head-node, leaves are body-nodes maximum path length 1.

Figure 8.4 : Features, their symbols and properties in clausal models

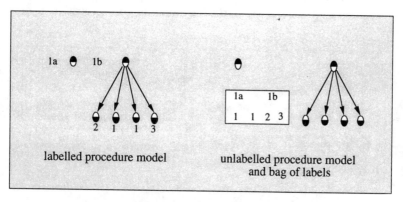

labelled procedure model

unlabelled procedure model
and bag of labels

Figure 8.5 : Procedure models

removes two types of ordering: the ordering of the body predicates, and the ordering of the trees themselves. These models, too, while initially labelled, have their labels removed and 'bagged', as for clauses. Each tree of the procedural forest model has the features of Figure 8.4.

8.2.2 Modelling compositions

Each of the four composition operations for Prolog – conjunction, clausal implication, disjunction and hierarchical implication – must be represented in the structural model.

Modelling conjunction

The conjunction operator in Prolog is the comma. It groups separate predicates into a sequence within a clause body. Our model does not retain ordering of predicates in a clause body, therefore the conjunction operator is merely a grouping device. We show it on the model by representing the conjoined body-nodes as a single digraph, albeit disconnected. This fairly trivial process is shown in Figure 8.6.

Modelling clausal implication

Clausal implication links the head of a clause to its body as in

$$head :- pred_1, \dots pred_n$$

which reads "head is implied by $pred_1$, and ... and $pred_n$".

Modelling this implication by a group of arcs leading from the head to each of the body predicates leads to the model of Figure 8.7.

Modelling disjunction

In Prolog, clauses with matching heads are grouped by disjunction, and the result is called a procedure. Similarly, the modelling of a procedure is brought

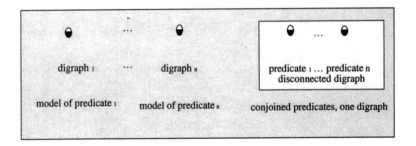

Figure 8.6: Modelling conjunction

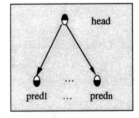

Figure 8.7: Modelling clausal implication

about by the grouping of individual clause models (Fig. 8.8). This model of disjunction uses the *labelled* clause models to produce a *labelled* procedural model. The labels may then be further abstracted as usual.

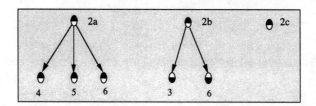

Figure 8.8: Modelling disjunction as a grouping of clause models

Modelling hierarchical implication

Hierarchical implication is modelled by one or more arcs from body-node to the one or more matching head nodes of a lower-level procedure model. Referring to the example program of Figure 8.1, Figure 8.9a shows such an arc between node 2, modelling a body predicate, and each of the nodes 2a, 2b and 2c, which are models of the clauses of procedure 2. Once again this model of composition by hierarchical implication is carried out on the labelled procedural models, and produces a labelled hierarchical model.

Hierarchical composition also links the start node to the top level procedure of a program. Figure 8.9b shows the start node, hierarchically composed with the top level procedure of the model of Figure 8.9a.

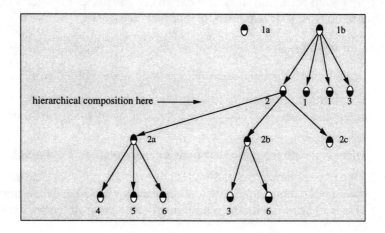

Figure 8.9a: Modelling hierarchical implication

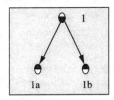

Figure 8.9b: Modelling hierarchical implication on the start node

8.3 Modelling programs

Modelling of programs is presented in two stages: first, a 'basic' program model which retains all the modelling features just discussed, and then a 'condensed' program model which is an abstraction of the basic one.

8.3.1 The basic program model

For convenience, we divide programs into two classes: programs which contain no recursion, and programs with recursion.

The basic model of a program with no recursion

As we have seen, programs are constructed by hierarchical implication of procedures. In the same way, we build program models by hierarchically composing their procedure models. Figure 8.10 shows code, stripped text and program model for a short, but non-trivial, program, 'apartment' (3), which does not contain recursion. It is also extremely simple, as disjunctions only occur at the lowest program level, that of fact clauses.

Features of the digraph are as follows.

- Node 1 is the start node: the unique body-node with no entry arc.
- The digraph is bipartite, linking head-node to body-node by clausal implication arcs, and body-node to head-node by hierarchical implication arcs.
- There is a path from the start node to every node in the digraph.
- All paths terminate in head-nodes corresponding to fact-clauses, or to body-nodes representing system predicates whose heads are implicit. This is in accord with the requirement that a program be resolvable to fact clauses.

For such a program, the digraph model is a tree.

```
:-plan(FD,D1,W1,D2,W2)                              1. :-plan.

plan(FD,D1,W1,D2,W2):-   frontroom(FD,D1,W1),       1a.plan:-  2.frontroom,
                         opposite(D1,D2),                      5.opposite,
                         room(D2,W2),                          3.room,
                         notopposite(W1,W2).                   6.notopposite.
frontroom(FD,D,W):- room(D,W),                      2a.frontroom:- 3.room,
                    direction(FD),                                 4.direction,
                    FD=\=D,FD=\=W.                                 =\=,=\=.
room(D,W):-   direction(D),                          3a.room:-     4.direction,
              direction(W),                                        4. direction,
              D=\=W,W=\=north.                                     =\=,=\=.
direction(north).                                   4a.direction.
direction(south).                                   4b.direction.
direction(east).                                    4c.direction.
direction(west).                                    4d.direction.
opposite(north,south).                              5a.opposite.
opposite(south,north).                              5b.opposite.
opposite(east,west).                                5c.opposite.
opposite(west,east).                                5d.opposite.
notopposite(D1,D2):- opposite(D1,D3),               6a.notopposite:- 5.opposite,
                     D2=\=D3.                                        =\=.
```

program text   **stripped text**

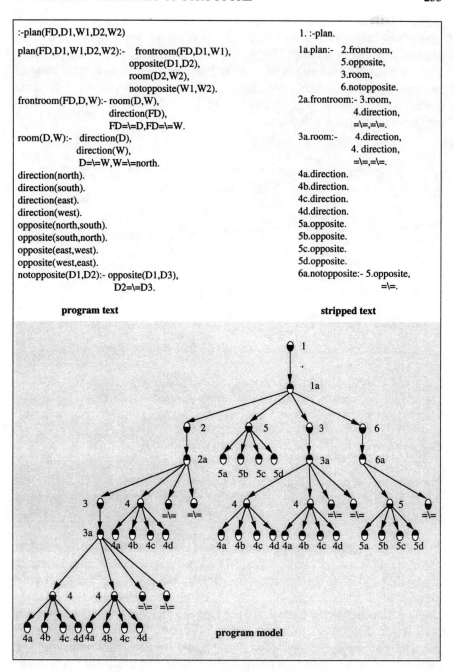

program model

Figure 8.10: Model of program 'apartment'.

In isolating the label bags of clause and procedure digraphs, we were able to distinguish between the root and leaf labels (there were no other) by their nature, suffixed or not. If we apply the same modelling step to the hierarchy of Figure 8.10, this is no longer true. To lose as little information as possible, we may partition the labels into three classes: the root node labels, the internal node labels, and the leaf labels. Figure 8.11 shows this partition, which suggests measures for Section 8.4.

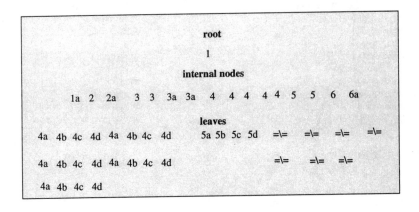

Figure 8.11: Labels of Figure 8.9 partitioned into 3 groups

The basic program model of a program with recursion

The example program of Figure 8.1 contains several instances of recursion. We use it to illustrate the extra features that recursion introduces to the basic program model.

Consider the hierarchical composition of procedure 'quicksort' and procedure 'split', as illustrated in Figure 8.12a. Node 1 is the start node. Nodes 1a and 1b model the head-nodes of procedure 'quicksort', and nodes 2a, 2b and 2c model the head predicates of procedure 'split'. Confining ourselves at this point to the single subtree rooted at node 2, we see that hierarchical composition on that node reintroduces an identical node at a lower level. This is a consequence of the recursive nature of procedure 'split', and hierarchical composition, if allowed, would proceed to infinity. This is clearly unacceptable in a model which must be finite.

The solution to the modelling problem is to allow each labelled node to appear only once in a hierarchical sequence, each subsequent appearance being

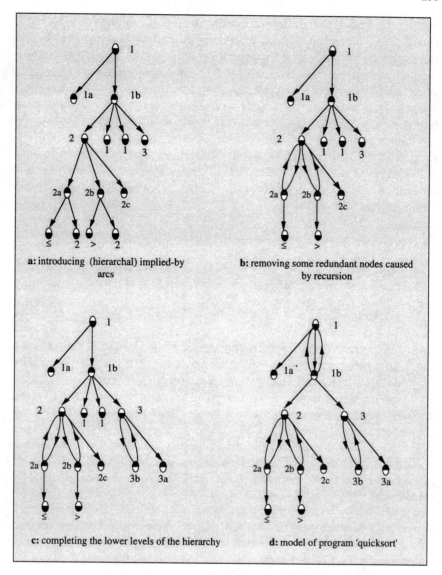

Figure 8.12: Development of program model for 'quicksort'

represented by an arc upwards to its first occurrence. Figure 8.12b illustrates this for this same node, 2, and localizes the recursive structure to a single cycle. The presence of recursive cycles destroys the tree-like properties which could otherwise result from hierarchical composition, but it means that, instead of an infinite tree, the program is modelled as a finite cyclic digraph. No loss of information is involved here since the new arc can be *unfolded* to its original position.

Adding the model of procedure 'append' to node 3, the digraph of Figure 8.12c is obtained. There is one final development to be made to the model. We need to merge the start node with the two other similar nodes labelled 1, linked by out-arcs of node 1b. Doing so achieves the full program model of Figure 8.12d.

Program 'quicksort' yields a model which, in addition to the features of the non-recursive program 'apartment', also illustrates how to represent recursion by finite models. Since the model of the recursive program is not a tree, it has no root. There is sometimes a difficulty in assigning a unique start node, since, in cases of recursion in the top-level clause, there may be paths to all other node as in Figure 8.12d from nodes 1 and 1b. Where this occurs, we designate a *start group* instead of a start node.

Recursion can be direct or indirect, and this is reflected in the model:

- Direct recursion (where a body predicate has the same name as the procedural head predicate) yields a directed graph model with one or more cycle of minimal size. Figure 8.12 includes instances of this on nodes 1, 2 and 3.

- 'Indirect recursion' (where a body predicate has the same name as a more remote ancestor) is modelled by a directed graph with cycles of non-minimal size. This type of recursion does not occur in 'quicksort, but it is demonstrated in the contrived program fragment, and its model in Figure 8.13.-

Both in the case of direct recursion and of indirect recursion, there may be more than one arc redirected to the same-label ancestor node. Figure 8.12d illustrates this for direct recursion on nodes 1 and 1b. This group is an example of two *fused* cycles.

Figure 8.14 summarises the symbols and properties of a labelled program model. Figure 8.15 draws attention to those features of the program model which are peculiar to recursion.

8.3.2 The condensed program model

The model of any realistic program is sure to be cumbersome and complicated. We therefore introduce a simplified form, developed by the progressive removal and recording of digraph features.

The program model is simplified by a process called *condensation*. This takes place in stages. Following our established procedure, the information removed is documented, and its meaning in the original object examined.

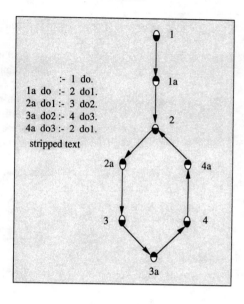

```
                :- 1 do.
      1a do  :- 2 do1.
      2a do1 :- 3 do2.
      3a do2 :- 4 do3.
      4a do3 :- 2 do1.

        stripped text
```

Figure 8.13: General example of indirect recursion.

feature	symbol	properties
head predicate	◑	one or more in-arcs; zero out-arcs it models a fact clause; else one or more out-arcs ;
start predicate or predicate in body of procedure	◕	zero in-arcs if unique start node; else one or more in-arcs; zero out-arcs if system predicate node else one or more out-arcs
'is-implied-by' relationship	↓	clausal or hierarchical implication, alternating ;
program	see e.g.figures 8.10 and 8.12d	in absence of recursion the bipartite digraph is a tree.

Figure 8.14: Features, their symbols and model properties in program models

feature	no recursion in program	recursion in program
bipartite digraph	tree	connected digraph containing cycles;
start node/s	single node, root of tree	as non-recursive if start procedure is non-recursive; else start group of nodes;
paths	no cycles; all paths terminate in leaf nodes which are head-nodes of fact clauses or system predicate body-nodes	paths may contain or terminate in cycles.

Figure 8.15: Comparison of models of recursive and non recursive programs

First stage condensation

The first stage of condensation is to fuse each source body node with its target head nodes, removing all self-loops generated in the process, and representing each fused node by the new symbol, •. In the program model of Figure 8.10, node 1 is a source body-node. Its target head-node is node 1a. These are fused to form a single node, labelled with the common integer label of constituents. Other groups of nodes to be fused are nodes 2 with 2a, 3 with 3a, 4 with 4a, 4b, 4c and 4d, etc.

This condensation abstracts some properties from the model. The remaining digraph – in this case a tree – is no longer bipartite. The abstracted properties are those of the internal structure of all nodes, together with their labels, and, in addition, the system predicate labels. For the case of program 'apartment' of Figure 8.10, the products of this process are shown in Figure 8.16. The abstractions of Figure 8.16 are called *simple primes*, shown beneath the *first stage condensed digraph* in Figure 8.16. In a further modelling step, the labels of the system predicate nodes are removed.

Carrying out the same condensation process on the model of program 'quicksort' gives the digraph and simple primes of Figure 8.17. The condensed digraph model of this program is now a tree, with a single start node.

Note the difference between procedural models and simple primes.

- There are no cycles in procedural models; to form a cycle requires hierarchical implication, which is absent from the procedure model. Simple primes may contain cycles.

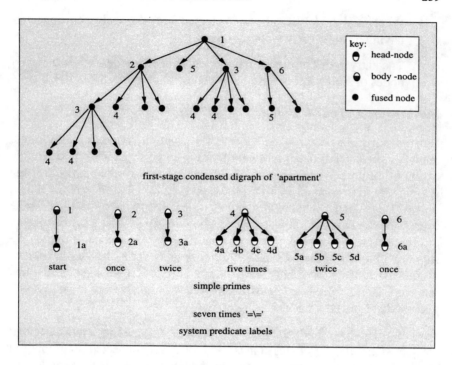

Figure 8.16: First stage condensed models of program 'apartment' - tree and its primes

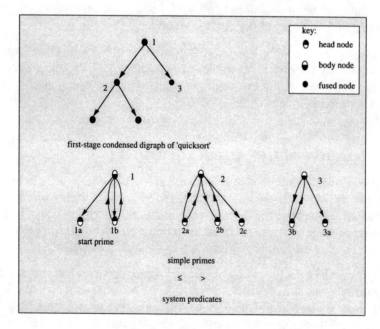

Figure 8.17: The first stage condensed digraph of program 'quicksort' and its primes

- Procedural models may be unconnected; primes are always connected, having a single entry node.

- The entry nodes of procedural models are head-nodes, while the entry nodes of simple primes are body-nodes, see Figures 8.8 and 8.17.

Further stages of condensation

When carried out on the basic model of a program containing only direct recursion, first stage condensation will lead to a tree structure, because of the fusion of the source and target nodes of self-loops into a single node. If the original program contains indirect recursion such as that of Figure 8.13, then the first stage condensation will remove any self-loops, but will still leave the cycles of indirect recursion. For a model containing a subdigraph, such as that of Figure 8.18, a second stage of condensation will be needed to reduce the structure to a tree. If there exist yet further nested cycles, then further stages of condensation are carried out, until the structure reduces to a tree. Structures removed during second and subsequent stages of condensation are 'knots': maximally strongly connected subgraphs.

These further stages of condensation have been defined mathematically and are quoted in full in ([1]). Nodes of maximally strongly connected subgraphs are *condensed* to a single node. This results in formation of self-loops and these are subsequently enclosed in the single node. In the process, in-arcs of any of the group become in-arcs of the condensed node, and out-arcs of any of the group become out-arcs of the condensed node.

The primes removed during second stage condensation are called 'complex primes'. Figure 8.19 shows examples of some other possible complex primes, though structures c, d and e of that figure should not be encountered in Prolog programming practice.

8.4 Measures of the structural model

The models of the previous section reveal properties of the model which can be captured in scalar counting measures. One may then examine what attributes these represent in the original object – the Prolog logic text. It is clear that there are two components in all the models produced: the digraph and its labels. Together they serve to characterise the object modelled.

Measures on a collection of labels are the same, however large or small the program or program fragment, and irrespective of the origin of the labels. Figure 8.20 shows a group of counting measures on a set of labels. To ascribe these to a particular originating entity, it is sufficient to add a distinguishing suffix such as 'p' for procedure or 'c' for clause.

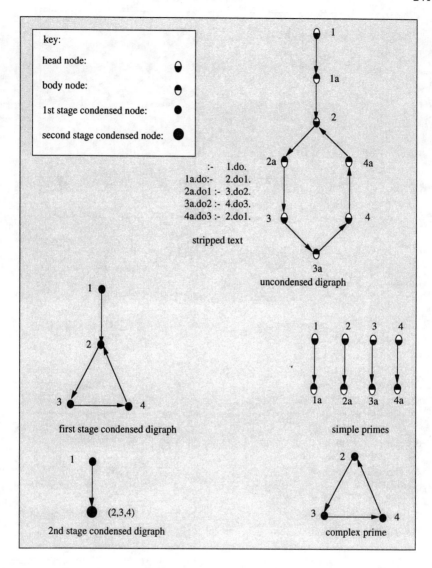

Figure 8.18: First and second stage condensation, showing abstraction of a complex prime from a first-stage condensed digraph model

8.4.1 Measures on the clausal model

In addition to the items of Figure 8.20, the clausal digraph yields the measures shown in Figure 8.21. The model reveals, as a countable feature, the number of out-arcs from the entry node of the clause. This is the same as the number of leaves.

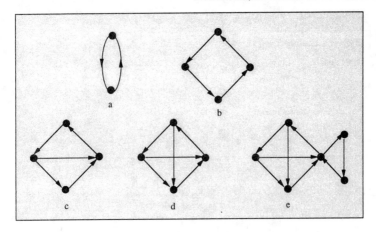

Figure 8.19: Other, possible complex primes

countable feature	parameter symbol and derivation	Prolog feature
1. cardinality of bag of body labels	m	no. of body predicates
2. cardinality of bag of system predicate symbols	p	no. of body system predicates
3. cardinality of bag of user label symbols	u also derivable as m - p	no. of program defined body predicates
4. cardinality of set of body labels	n	no. of distinguishable body predicates in clause
5. cardinality of set of system predicate symbols	q	no. of distinguishable system predicates in clause
6. cardinality of set of user label symbols	v also derivable as n - q	no. of distinguishable programmer-defined procedures in clause body
7. number of repetitions of head label name among body predicates	r	extent of recursion

Figure 8.20: Countable features of a collection of labels taken from structural models up to condensation

8.4.2 Measures on the procedure model

The labelled and unlabelled models for a procedure with only one clause are indistinguishable from a clause model. If the procedure model is disjunctively composed of more than one clause, then the procedural model is composed of a group of clause models. The measures it generates can either be left as a bag of each, or summed over the procedure. The main extra measure generated will be the number of clauses making up the procedure. Measures, in addition to those of Figure 8.20, are shown in Figure 8.22.

countable feature	symbol	meaning in Prolog text
1. no. of out-arcs on root	l_c **	no. of body predicates
2 no.of leaves of tree *	l_c **	no. of body predicates

* numerically the same as 1   ** lower case L.

Figure 8.21: Countable features of unlabelled clausal tree

countable feature	symbol, name and/or derivation	meaning in Prolog text
1. no. of out-arcs of trees of the forest	summed as l_p **	no. of body predicates in procedure
2 no. of leaves of trees of the forest *	summed as l_p **	no. of body predicates in procedure
3. number of trees in procedural forest	extension, e_p	no. of clauses in procedure

* numerically the same as 1   ** lower case L

Figure 8.22: Countable features of unlabelled procedural forest.

8.4.3 Measures on the program model

Overall measures on the label bag

Any pre-condensation measures derived from the label bag may be partitioned as in the example of Figure 8.11, but otherwise they are very similar to those of Figure 8.20.

First-stage condensation is carried out on the labelled program digraph. These labels will be part of the collection of labels attached to simple primes. Measures of Figure 8.20 apply.

Measures of hierarchical structure

As we have just seen, condensation systematically removes the primes, such that the ultimate outcome of the condensation process is a tree, revealing the backbone of the hierarchical structure. To measure the characteristics of the structural model, a measurement scheme is needed, comprising:

- the prime measures which characterize condensed nodes,
- measures of the tree.

To capture the distinction between simple and complex primes, we need:

- nodal measures of the simple primes,
- nodal measures of the complex primes.

Measures of simple primes

Figure 8.23 shows some of the simple primes often encountered in practice. Each simple prime has a single-entry body-node. No node is more than unit distance from the entry node, and each node is an entry node, a leaf node, or a member of a cycle.

The salient features of these primes may be captured in three measures, named *extension, cyclic_order* and *implication_list.*

- 'Extension' is defined as the number of out arcs from the entry node.

- 'Cyclic_order' is the number of independent cyclic subgraphs in the prime: subgraphs with no common arc.

- 'Implication_list' is a list of the number of paths in each of the independent cyclic subgraphs of the prime. For an acyclic prime the list is empty [ ], and the number of items in the list equals the cyclic order of the prime. An individual item in the list is called an 'implication measure'.

The measures are summarised in Figure 8.24, and those of the primes of programs 'apartment' and 'quicksort', shown in Figure 8.23, are given as illustration in Figure 8.25.

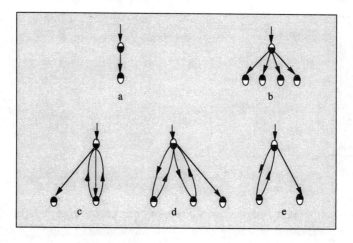

Figure 8.23: Some simple primes

countable digraph feature	parameter name and symbol	Prolog feature
number of out-arcs on entry node	extension e_s	number of dependent body predicates in the procedure
number of cycles	cyclic order c_s	number of recursive clauses in a procedure
fused cycles	implication list i_s	list of number of recursive calls per clause

Figure 8.24: Countable features of simple primes

A ranking order may be created for simple primes which reflects some measure of their intricacy. (We deliberately avoid the contraversial word 'complexity'.) Implication may be thought of as the most appropriate indicator. Within the same implication_list, intricacy is ranked by cyclic order, and within the same cyclic order, ranking is by extension.

Measures of complex primes

Complex primes can contain any number of nodes greater than 1. It is conceivable, though in practice unlikely, that these will be more than just simple cycles. All eventualities are catered for in the proposed measures for complex primes. There are three of these:

parameter	a	b	c	d	e
extension e_s	1	4	2	3	2
cyclic order c_s	0	0	1	2	1
implication list i_s	[]	[]	[2]	[1,1]	[1]

Figure 8.25: Measures of the simple primes of Figure 8.23

- 'Scope' is defined as the number of nodes in the complex prime (or knot or maximal strongly connected subgraphs). Figure 8.19e shows a complex prime whose scope is 6.

- 'Cyclic order' is, as before, the number of independent cyclic subgraphs in the prime. In Figure 8.19e its value is 2.

- As in the case of simple primes, 'Implication list' is a list of 'implication measures'. The number of items in the list is given by the cyclic order. Each implication measure gives the number of complete paths in a fused cycle. For a knot, the lowest value of this measure is [1], indicating: a list comprising a single item whose implication measure is 1. Again using Figure 8.19e as the example, the left-hand cycle has 3 paths and the right-hand one has 1. The implication list is thus [3,1].

These measures are summarised in Figure 8.26, and the measures for the complex primes of Figure 8.19 are in Figure 8.27.

Just as for simple primes, a ranking order may be imposed on complex primes. Here again, the implication list is the main guide, but scope is used as a further feature, to distinguish the several graphs whose implication list is the same (such as the graphs referenced in the first three columns of Figure 8.27).

Structural measures on the condensed tree

The tree measures described in Chapter 3, including size, width and depth, are directly applicable as the third component of the measurement scheme for the structural model, and we do not elaborate on them further.

8.5 The measurement tool SPA

A first tool for structural measurement of Prolog, the Static Prolog Analyser SPA, was part of a set of modules [4] created for the measurement of Prolog.

digraph feature	parameter name and symbol	Prolog feature
number of nodes in the complex prime	scope s_m	size of recursive unit
number of independent cycles in the digraph	cyclic order c_m	number of different multiply recursive groups in the knot
number of paths through the digraph	implication i_m	related to difficulty of comprehending the code

Figure 8.26: Countable features of complex prime digraph

parameter	fig 8.19a	fig 8.19b	fig 8.19c	fig 8.19d	fig 8.19e
scope	2	4	4	4	6
cyclic order	1	1	1	1	2
implication list	[1]	[1]	[2]	[3]	[3,1]

Figure 8.27: Some complex prime measures

Module 1 generates stripped text. From this, Module 2 isolates simple primes, Module 3 the complex primes, and Module 4 the tree structure. (Module 5 of this early form of an automated tool also included a simple version for the model-based measurement of data, and is the basis on which the tool PDQ, referred to in Chapter 10, was subsequently built). The tool SPA now needs updating and a further module added, to produce the tree measures defined in Chapter 3.

8.6 Summary

This chapter offers a method of model-based measurement of the structure of the Prolog logic text. The model represents the logic text as a directed graph, showing the underlying tree, the simple primes and the complex primes (the strongly connected graphs) which are nested in the nodes of the tree. The measures given here reveal the graph properties of the primes.

Recall the general expression of system structure $M_s(P(t)) = (\mathbf{Comp}, \mathbf{R}_{Comp})$, first presented in Chapter 2. This is the basis of a general method for measuring product design, proposed in Section 5.2.6. For the present case,

> **Comp** is the set of simple and complex primes yielding set measures and individual prime measures. These are the measures presented in this chapter.
>
> **R**$_{Comp}$ is the tree which forms the backbone of the structure of the Prolog logic text. The modelling procedure which uncovers the tree is given in the chapter, and the tree measures in Chapter 3 serve to characterize it.

Refering now to the general measurement scheme in Figure 3.8 of Chapter 3, the measurement scheme for the Prolog logic text is built on the individual prime measures and tree measures which are the (direct and indirect) measures of the scheme. These together give the object-oriented characterization of the structure of the Prolog logic text. It is then a matter of judgement to decide how the tree and nodal measures should be combined to form a utility measure for the structure over-all. The design of a utility measure is beyond the scope of this book, but the reader may note the analogy with the development of structural measures in procedural programs (see e.g. [5].)

8.7 References

1   Myers M (1987): MSc dissertation, "The introduction of axiomatic theory of structures to Prolog" South Bank University.

2   Clocksin W F, Mellish C S (1984): "Programming in Prolog". Springer-Verlag.

3   Coelho H, Cotta J S, Pereira L M (1982): "How to solve it with Prolog", Laboratorio Nacional de Engenharia Civil, Lisbon.

4   Benwood H J J (1988): "A software tool for the structural measurement of Prolog", M.Sc dissertation, Heriot-Watt University.

5   Fenton N E (1991): "Software metrics". Chapman and Hall.

9 MODELS AND MEASURES OF DATA

9.1 Introduction

This chapter demonstrates a model-based measurement scheme at work on the data of the Prolog text. The abstraction of the logic component from the Prolog text is described in Section 7.3. Figure 7.6 shows that, for the purpose of modelling and measurement, the Prolog logic text divides into two parts: structure over data and the hierarchical structure of the logic text. This chapter is devoted to the former ([1]).

The direct measures of Prolog data arise naturally from a graph model developed in Section 9.2. The definitions of measures are given in Section 9.3, and are illustrated on an example in Section 9.4. Section 9.5 summarizes the emergent measurement scheme and indicates its uses.

9.2 Models of the data of Prolog logic text

9.2.1 Models of the data entities

The Prolog logic text is derived from the Prolog text by the series of abstractions shown in Section 7.3. The data is represented with the aid of directed graphs. The position of the data in relation to the structure of the program, the data objects of the head or body of a Prolog clause, and the structure of the terms, are all retained, but the model only partially preserves order. It suppresses the order of clauses in a procedure and the order of predicates in the body of a clause, but retains the order of data objects within the predicates. Preserving the head/body relation allows the representation of the implication relationship between the two.

Symbols of the directed graph model

The digraph of Prolog data is composed of three main types of node and one kind of arc.

- *The nodes* of the directed graph represent arguments, atomic terms and functor symbols. The graph is bipartite in the sense that: a) each path originates in an argument node and terminates in an atomic term, and that b) between the two, functor symbol nodes and argument nodes alternate.

  * *Arguments* are nodes represented as squares. They have in-degree 0 or 1 and out-degree 1 or more.

  * *Atomic terms* are represented as circles. A plain circle stands for a variable in the code. A circle containing the letter C models a constant. All atomic entities retain their labels up to clause construction. 'Don't care' data types are variables whose label is the symbol '_'. All these nodes have out-degree 0. On removal of variable labels at clause construction, the variable data type nodes are further distinguished, remaining as plain circles for the first occurrence of a variable, and changing to a circle containing an '=' sign for a variable repetition.

  * *Functor symbols* are represented by unlabelled *structure bars*, nodes having in-degree 1 (up to modelling disjunction) and out-degree 1 or more.

- A *directed arc* models the 'contains' relationship which binds together the nodes of the graph. (Bear in mind that the direction of arrows in this and all other data digraphs does *not* imply direction of data flow.)

The symbols and their names are shown in Figure 9.1

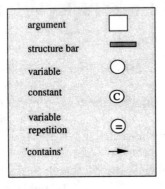

Figure 9.1: Symbols of the directed graph models of data

The hierarchy of Prolog constructs

This section presents the models of the data of each of the composite Prolog language entities, composed as directed graphs. The entities are listed from the simplest (compound term) to the most elaborate (the complete Prolog program), and are given in the series of Figures 9.2a to 9.2n. The method of construction is detailed later, in Section 9.2.2. The generic terms 'data tree', 'data forest' and 'data digraph' are applied throughout.

- Each *compound term* is modelled as part of a bipartite data tree of arguments and terms. The entry node is the structure bar modelling the main functor. The structure bar will in turn generate one or more new arguments, each of which may contain an atomic term or a compound term (Figure 9.2a).

- A *data object (argument with term)* is modelled as a bipartite tree whose root is an argument, from which the single arc leads to the main structure bar of a compound term model or to the model of an atomic term, whichever is the content. An example of this last is given in Figure 9.2b.

- A *predicate* is modelled as a data forest, consisting of individual data trees whose roots are argument nodes, Figure 9.2c.

- A *clause head* is modelled as a data forest, enclosed in a frame. The trees of the forest are root-uppermost and the arcs are down-directed, shown in Figure 9.2d. The frame represents the boundary of the variable scope. (The terms *up* and *down* always refer to position on the page.)

- A *clause body* is modelled as a data forest enclosed in a frame whose trees are root-down and whose arcs are up-directed, Figure 9.2e. The conjunctive composition of predicate data forests into clause body data forests is described in Section 9.2.2.

- The *start predicate*, which is an isolated, single-predicate clause body, is likewise modelled as a data forest, enclosed in a frame whose trees are root-down and whose arcs are up-directed, Figure 9.2f.

- A fully labelled *fact-clause* model is identical to the model of a clause head, Figure 9.2g. An abstraction on this model is then produced. The simplified fact-clause model is a partially labelled data forest. It is derived from the fully labelled model by replacing every variable repetition with a node carrying the 'identity' sign (=), and then removing all variable labels. The rightmost of a pair of identically labelled nodes is considered to be the repetition; Figures 9.2h and 9.2i illustrate the necessity for this constraint (see also Notes). All labelling other than identity sign is then omitted. Omission of variable labels removes the necessity for a scope frame. The simplification sacrifices the ability to recognize which variables are being repeated, but retains the information needed to count of 'bag of variables' and 'set of variables'.

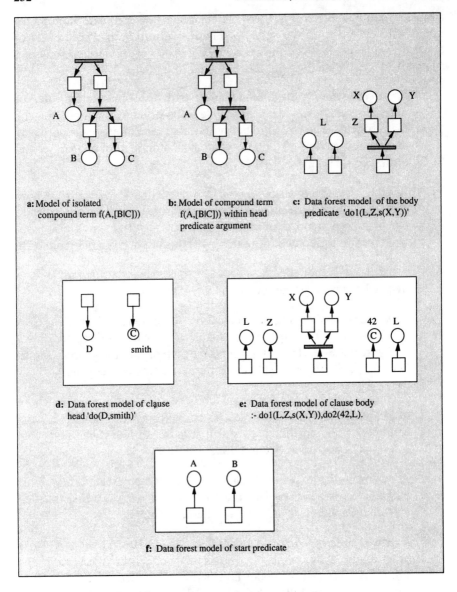

a: Model of isolated
   compound term f(A,[B|C]))

b: Model of compound term
   f(A,[B|C])) within head
   predicate argument

c: Data forest model of the body
   predicate 'do1(L,Z,s(X,Y))'

d: Data forest model of clause
   head 'do(D,smith)'

e: Data forest model of clause body
   :- do1(L,Z,s(X,Y)),do2(42,L).

f: Data forest model of start predicate

Figure 9.2: Models of Prolog data entities (Contd. over)

- In a fully labelled *rule-clause* model the head and body data forests are
  fused at the leaves of corresponding name, as illustrated in Figure 9.2j.
  The abstraction of a simplified rule-clause model is derived from this as
  in the case of the fact-clause and illustrated in Figure 9.2k. The fusing
  converts the two data forests to a data digraph. Argument nodes at the
  top of the digraph represent arguments in the head of the rule-clause;

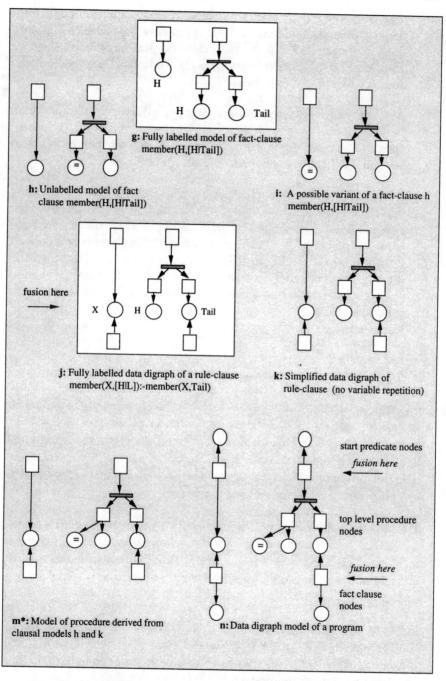

g: Fully labelled model of fact-clause
   member(H,[H|Tail])

h: Unlabelled model of fact
   clause member(H,[H|Tail])

i: A possible variant of a fact-clause h
   member(H,[H|Tail])

fusion here

j: Fully labelled data digraph of a rule-clause
   member(X,[H|L]):-member(X,Tail)

k: Simplified data digraph of
   rule-clause (no variable repetition)

start predicate nodes
fusion here

top level procedure
nodes

fusion here

fact clause
nodes

m*: Model of procedure derived from
clausal models h and k

n: Data digraph model of a program

* l omitted for typographic reasons.

Figure 9.2 (contd.): Models of Prolog data entities

argument nodes at the bottom represent arguments in the body. Target nodes representing atomic variable types may now have two in-arcs. Such in-arcs are either down-directed if originating from the head, or up-directed if originating from the body. The partially labelled model does not require a scope frame.

- A Prolog *procedure* is modelled as the minimal bipartite digraph which contains as subdigraphs all the clausal digraphs comprising the procedure. Instructions for constructing such a digraph are included in Section 9.2.2 and a simple example shown in Figure 9.2m.

- A Prolog *program* is modelled as a hierarchy of Prolog procedure models linked through source arguments, starting with the model of the start predicate. The links are established by head predicate and body predicate names acting as labels for groups of arguments and derived from the initial code. Figure 9.2n shows a program model as a bipartite digraph. At the top are the term and argument nodes representing the start predicate of the program. At the lower end of the digraph are subgraphs which are trees. Other features of the program model are discussed later.

Notes to Figure 9.2

Figure 9.2g, containing variable identifiers, requires a scope frame, while Figure 9.2h does not. Linking the identity label H to one of the variables, and interpreting the other as 'repetition', introduces non-uniqueness, as in Figures 9.2h and 9.2i. Uniqueness is restored if we restrict the 'repetition' label to the rightmost variable node. Nevertheless, both representations interpret the notion of 'a data structure with two independent variables with one repetition' eventually yielding a unique measure.

The data digraph of Figure 9.2k is unlabelled and does not need a scope frame. The directions of the arcs show which variables are contained in the head arguments and which in the body. This figure also indicates that variable nodes delineate the interface between head and body of a rule-clause.

Figure 9.2m is the procedure model of the clauses modelled in 9.2h and k. Both are subgraphs of the procedural digraph. The procedural digraph shows that argument nodes may now have more than one out-arc, in this case both are downward directed.

The procedure model of 9.2m is fused, at the top level of the hierarchy, to a model of the start predicate and to the model of a simple fact-clause 'do(X,Y)' at the bottom level. The argument nodes through which the levels interact are recognizable by having two out-arcs, one up and one down.

End of Notes

9.2.2 Modelling compositions

The four composition operators of the Prolog world are now mapped into the model world. Figure 9.3 is a general schema, showing the way in which a model represents the composition of textual Prolog entities by means of the basic Prolog constructors of conjunction, disjunction, clausal implication and hierarchical implication. The modelling process must be so defined that the model of the composite Prolog entity should be the same, irrespective of whether

- the two Prolog elements are composed first and the composite Prolog entity is mapped into its model, or

- the Prolog elements and the Prolog constructors are modelled first and then combined to form the model of the composite.

The four composition operators for textual Prolog were listed in Section 7.4.1 and illustrated in Figure 7.5a to 7.5c. They are now mapped into the models described above, using a 2-level program familiar to Prolog programmers:

EXAMPLE PROGRAM 'union/3'

```
union([],Set,Set).
union([Element|Rest],Set,Union):-
            member(Element,Set),
            union(Rest,Set,Union).
union([Element|Rest],Set,[Element|RestUnion]):-
            union(Rest,Set,RestUnion).
member(H,[H|Tail]).
member(Element,[H|Tail]):- member(Element,Tail).
```

END OF EXAMPLE PROGRAM 'union/3'

Modelling conjunction

In Prolog, predicates are conjoined by sequencing them and enclosing their variables in the same scope. The process is modelled by enclosing the predicate data forests in the same scope frame. Figure 9.4a illustrates conjunction, using the body predicates of the second clause of procedure 'union'/3 in the example program of the same name.

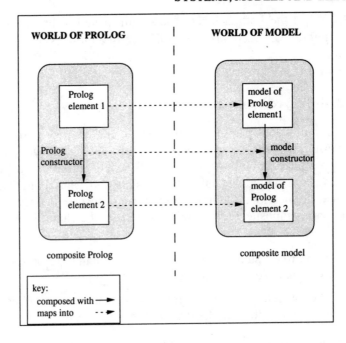

Figure 9.3: The schematic representation of the modelling of compositions

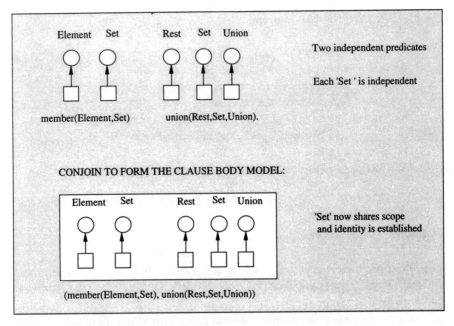

Figure 9.4a: Illustrating conjunction on body predicates

Modelling clausal implication

The data digraph model shows clausal implication by the relative position and orientation of the head forest and the body forest and the fusion of corresponding head/body nodes. The implication constructor gives a common frame for the head and body, locates them correctly at the top and bottom of the frame, and fuses the first occurrence of each leaf node in the head and body bearing a common label. The fusion models the communication channel between head and body, implemented in Prolog by the shared data labels. Figure 9.4b illustrates the fusion of the head forest and the body forest of the third clause of example program 'union/3' into a labelled clausal data digraph. Any remaining repetitions of variable label in the *fused* model are due to repetitions in the original head and/or to repetitions in the original body – one occurrence of Set and RestUnion but two occurrences of Element. The name 'head' is retained for the arguments and terms at the top of the frame, and 'body' for those at the bottom. The site of fusion is identifiable by the presence of variable atomic term nodes with in-degree 2.

Modelling disjunction

Prolog procedures are formed by disjunction of clauses whose heads match. A procedural data digraph model is a minimum digraph of which each of the data digraphs of the matching clauses is a subdigraph. In simple cases the procedural model is easy to produce as exemplified in Figure 9.2m. In more complicated cases it is useful to have a standard procedure for the construction of such a *super* digraph of the individual clausal digraphs.

The method is illustrated on example program 'union'. Figure 9.5 shows the data digraphs of each clause and stage in the disjunctive composition.

A METHOD

1. Choose the clausal digraph with the largest number of body 0 in-arc argument nodes. Call this the 'partial composite' (Figure 9.5a, chosen from 9.5a, 9.5b and 9.5d).

2. Until no more clausal digraphs are left:

    2.1   From the remainder, choose the clausal digraph with the most nodes.

    2.2   If the chosen digraph has more structure bars than the partial composite, then add such structure bars and their associated argument nodes to the partial composite.(One on cycle 1.)

    2.4   Repeat for constant nodes. (None on cycle one, one on cycle two.)

    2.5   Repeat for first occurrence variable nodes. (None in example.)

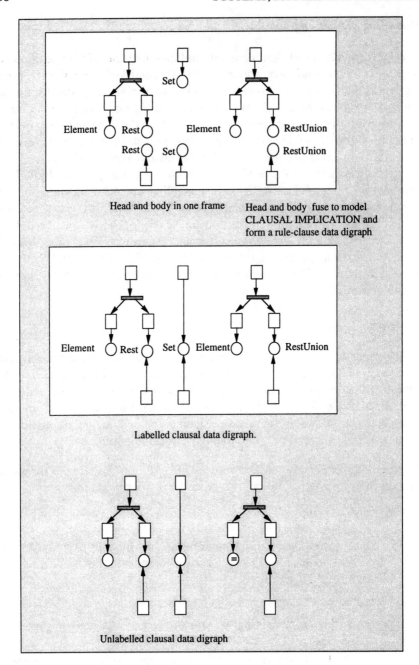

Head and body in one frame

Head and body fuse to model
CLAUSAL IMPLICATION and
form a rule-clause data digraph

Labelled clausal data digraph.

Unlabelled clausal data digraph

Figure 9.4b: Modelling clausal implication

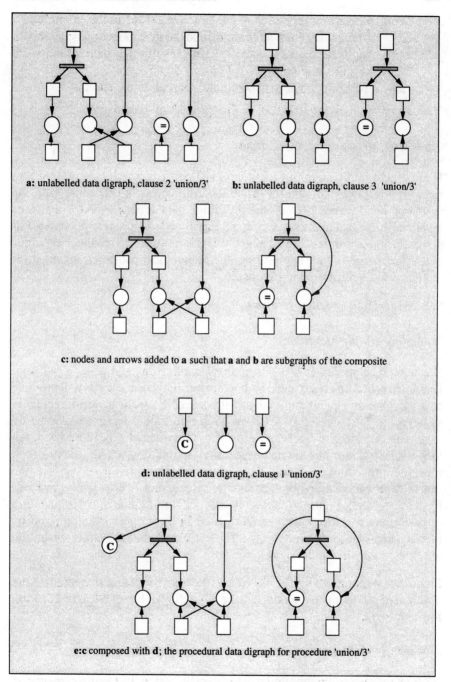

a: unlabelled data digraph, clause 2 'union/3'

b: unlabelled data digraph, clause 3 'union/3'

c: nodes and arrows added to **a** such that **a** and **b** are subgraphs of the composite

d: unlabelled data digraph, clause 1 'union/3'

e:c composed with **d**; the procedural data digraph for procedure 'union/3'

Figure 9.5: Unlabelled clausal digraphs of the three clauses of 'union/3'.

2.6   Repeat for repetition variable nodes. (None in the example.)

2.7   Proceeding from left to right, add the minimum arcs necessary to replicate the new clausal digraph within the partial composite (Figure 9.5c on cycle 1, Figure 9.5e on cycle 2).

2.8   Rename the new structure the 'partial composite'.

3. Rename the partial composite the 'procedural data digraph'.

Modelling hierarchical implication

Hierarchical implication in Prolog is modelled by hierarchical composition, merging the lowest arguments in the data digraph of a procedure with the matching head arguments of other procedure data digraphs modelling lower levels in the text. Labels present in the initial code as procedure names will select the correct procedure head and match arguments. The matching arguments are merged. The simplest possible hierarchical construct, invented for the purpose, is illustrated in Figure 9.6.

9.2.3 Modelling residual features

Modelling simple recursion

Recursion is a special case of hierarchical implication, and poses a problem. Linking a procedure head to the body predicate of the same name will produce a non-terminating hierarchical composition. (For the cyclic nature of recursion see Figures 7.4c and 7.5b). We have chosen to model recursive clauses by assuming an 'identity' fact-clause for the 'same-name' procedure head. An identity fact-clause has the name 'identity' and an arity equal to that of the recursive procedure. Each argument of the identity fact-clause contains the 'don't care' variable, the understroke '_'. Recursion is effectively 'pruned' from the digraph. Figure 9.7 shows the modelling of hierarchical implication of the recursive procedure 'member', used in the program 'union', and its identity fact-clause, 'identity(_,_)'. The individual data digraphs are on the left, the composite on the right.

Figure 9.7 illustrates both the use of the identity fact-clause to overcome the problem of modelling recursion, and the modelling of hierarchical implication for member/2.

Modelling indirect recursion

We treat indirect recursion similarly to direct recursion. Just as direct recursion is pruned by replacing the recursive body predicate by the equivalent identity fact clause, indirect recursion is pruned by replacing each indirectly recursive body predicate by its identity clause equivalent (Figure 9.8).

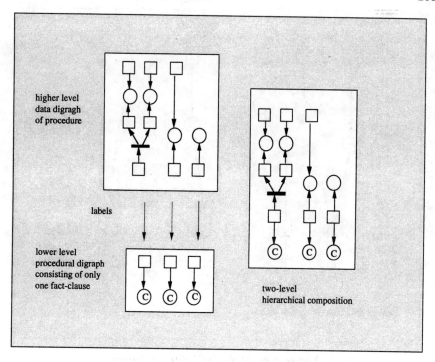

higher level
data digragh
of procedure

labels

lower level
procedural digraph
consisting of only
one fact-clause

two-level
hierarchical composition

Figure 9.6: Modelling hierarchical implication

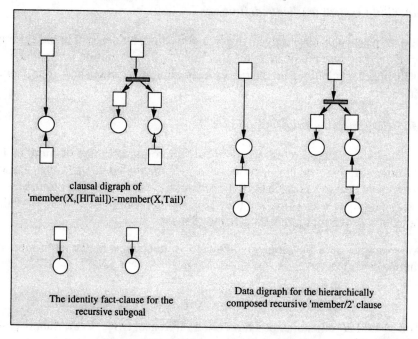

clausal digraph of
'member(X,[H|Tail]):-member(X,Tail)'

The identity fact-clause for the
recursive subgoal

Data digraph for the hierarchically
composed recursive 'member/2' clause

Figure 9.7: Modelling direct recursion

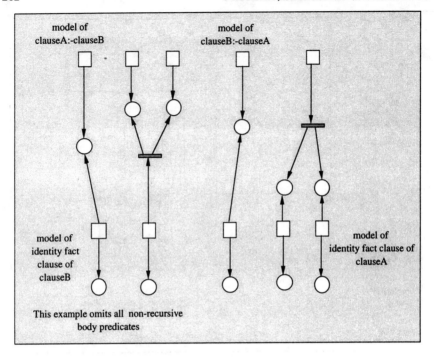

Figure 9.8: Modelling indirect recursion

Modelling system predicate definitions

System predicates have only an implicit definition, hidden within the system. Where the system predicate is retained, its implicit definition is modelled as its identity fact clause, with the requisite number of argument and 'don't-care' variable nodes.

Modelling metalogical predicates

A metalogical clause refers to complete subprograms in one or more of its arguments, rather than the terms we have so far modelled. Metalogical clauses are outside Prolog's domain of first-order logic, and thus beyond our model of the Prolog logic text. However, they are used in practice, and we explore the extension of the measurement scheme to them.

An example of a metalogical predicate is contained in the definition of 'intersect':

```
intersect(Xs,Ys,Zs):-
    setof(X,(member(X,Xs),member(Y,Ys)),Zs).
```

The second argument is not a term but a reference, twice, to the subprogram 'member/2'.

Figure 9.9 shows that the current model cannot represent such a clause satisfactorily. Instead, the figure introduces a 'metamodel', in which whole programs can be embedded in a single argument of such a metalogical clause. The box depicting the second argument of 'setof/3' contains within it two copies of the program model of 'member/2', and, if we decide to disregard the box itself, the model remains a bipartite digraph. This model clearly shows the presence of variable, identity, constant and structure nodes. Only the number of arguments is ambiguous.

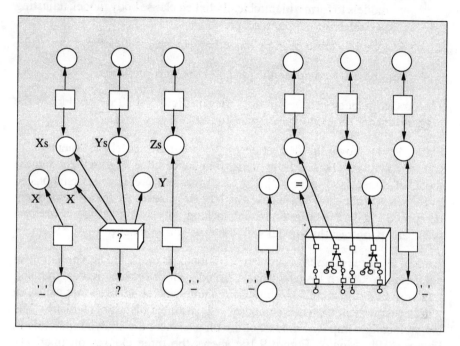

Figure 9.9: Partially labelled model of 'intersect/3' and a transparent view of the embedding argument

9.2.4 A procedure for modelling the data of programs

The application of constructors to the models in the production of a program model follows exactly the process outlined for Prolog in Figure 7.5a to 7.5c and thus fulfils the requirements stated at the start of this section and shown schematically in Figure 9.3.

To construct a program model

1. Model all individual clauses as in the preceding section.

2. Model groups of indirectly recursive clauses as in the preceding section.

3 Until top level procedure is reached:

   3.1 at the lowest level, disjunctively compose models of matching clauses to form procedure data digraphs, replace the clause models by the procedure model;

   3.2. link by hierarchical composition to next highest level body predicate models;

   3.3 conjunctively compose the hierarchically linked body predicate models to form a hierarchically linked clause body model, adjusting variables for common scope;

   3.4 join the clause head by modelling clausal implication.

4. Hierarchically compose the model of the start predicate.

The process is demonstrated in the series of Figures 9.10 which derives the program model for example program 'union/3'.

Figure 9.10a shows the derivation of the program model for 'member/2'. Clause 1 is a fact clause whose model first appeared as Figure 9.2h, and the model of clause 2 as Figure 9.2j. The recursive body predicate is first replaced by the identity clause (a fact-clause), and the two clauses are then disjunctively composed to give the program model, lacking only the modelling of the start predicate. (Contrast this model with the *procedure* model of Figure 9.2m.)

Each of the three clauses of procedure 'union/3' is modelled as a hierarchical composition. Clause 1 is a fact clause. Clause 2 features hierarchical implication with both the member/2 composite model and an identity fact clause in place of recursion. Clause 3 composes only with the identity fact clause. The hierarchically composed models of the three clauses are shown separately in Figure 9.10b. Finally, Figure 9.10c shows the three clauses disjunctively composed, with the addition of the start predicate model.

The features which all program models have in common are:

- their topmost nodes represent the terms of the start predicate;
- the lowest levels in the program model represent fact-clauses whose models are data forests; thus the lowest levels contain, as subgraphs, the trees of these data forests;
- the root of each of the trees of these forests is an argument which is identifiable in the model as being the argument node nearest the bottom which has at least one upward directed out-arc, as shown in Figure 9.11.

A feature of complete program models is that, both at top and bottom levels, all nodes are atomic terms. This readily distinguishes them from procedural models and incomplete program models.

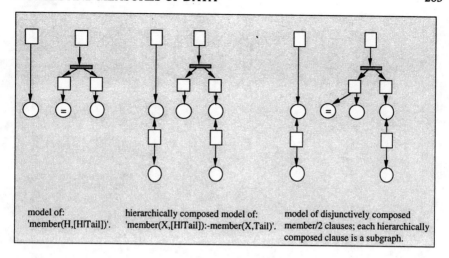

model of:
'member(H,[H|Tail])'.

hierarchically composed model of:
'member(X,[H|Tail]):-member(X,Tail)'.

model of disjunctively composed
member/2 clauses; each hierarchically
composed clause is a subgraph.

Figure 9.10a:
Derivation of model of program 'member/2' lacking only the start nodes.

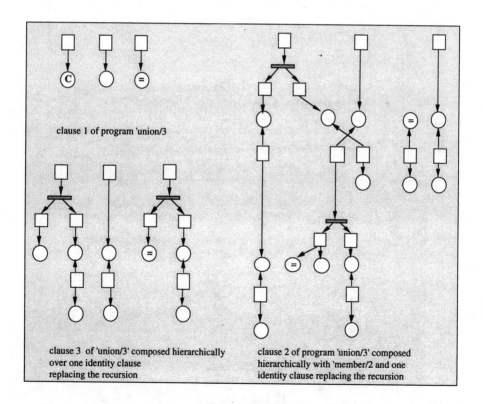

clause 1 of program 'union/3

clause 3 of 'union/3' composed hierarchically
over one identity clause
replacing the recursion

clause 2 of program 'union/3' composed
hierarchically with 'member/2 and one
identity clause replacing the recursion

Figure 9.10b: The three hierarchically composed models of the 'union/3' clauses

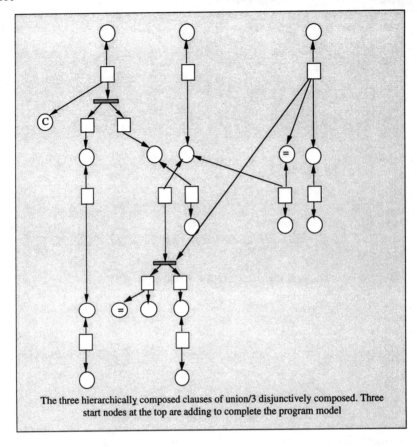

The three hierarchically composed clauses of union/3 disjunctively composed. Three
start nodes at the top are adding to complete the program model

Figure 9.10c: Complete model of program 'union/3'

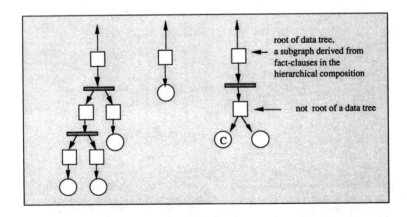

Figure 9.11:
Lowest level data subgraphs in a hypothetical program model composite.

9.3 Measures of Prolog data

In devising direct measures for Prolog data, six principles were adopted:

(i)   The measures should emerge from the model as their natural characterization.
(ii)   Measures should be independent, as far as possible.
(iii) Absolute (counting) measures should be used, if at all possible.
(iv)   The same measures should be used for characterizing the data of all Prolog objects.
(v)   The measures must have composition operators which accord with the compositions of Prolog, and with those of the models of Section 9.2.
(vi)   For the special purpose assigned to Prolog in Chapter 6, only features of the model which remain invariant over internal structure of the original Prolog should be used. The unfolded Prolog code should yield the same measures as a hierarchically structured version.

Section 9.3.1 identifies measurable features of the data model developed in Section 9.2. Section 9.3.2 selects from these a set of measures in accord with the above principles. These measures can characterize any segment of the Prolog logic text from the atomic (predicate) elements upwards, including complete programs and program segments with residual features such as recursion, system predicates and metalogical predicates. The measures may be read off the model, or, more conveniently, from the text itself, by composing predicate measures. The composition of measures is carried out in accord with an 'algebra' whose rules mirror Prolog syntax and the nature of the chosen measures.

9.3.1 Measurable features of data trees and data forests

As we have seen, a tree models each data object in a predicate. Qualitatively, these data trees are describable as labelled bipartite directed graphs, where

- the root is denoted by an argument symbol,
- all leaves are denoted by atomic term symbols,
- argument and term symbols alternate along all paths, see e.g. Figures 9.2b, 9.8.

Quantitatively such trees may be characterized by countable properties, such as those listed in Figure 9.12, in accord with the principles above.

The parameters of Figure 9.12 are not completely independent. The interpredicate relationships are summarized in the equations of Figure 9.13. The four parameters (S,V,I,C) of equation 9.7, together with D − the measure of the maximum depth of structure indicating the shape of the tree − can be used

countable parameters	chosen parameter name
no. of root argument nodes	A (= 1)
no. of argument nodes, both root and structure-generated	L
total no. of term nodes (structure bars and atomic data types)	T
total no. of leaves	Y
no. of constants (cardinality of constant bag)	C
no. of distinct variables (cardinality of variable set)	V
no. of variable occurrences (cardinality of variable bag)	X
no. of variable repetitions	I
no. of structure nodes (number of functors including lists, in Prolog text)	S
depth of nesting of structures (chosen measure: maximum no. of structure bar nodes between root and leaf)	D
out-degree of ith structure node (the nodal extension)	E_i

Figure 9.12: Countable parameters of the data tree

to characterize a data tree of binary structures. The correction factor ΔL supplements this for the general case where not all trees are binary. The measures of the data trees generalize to measures of data forests where $A>1$, measuring the number of trees in the forest as in Figure 9.14.

9.3.2 Definitions of measures of Prolog data entities

Six of the parameters of a data tree are now generalized for more than one tree, to form a 6-tuple $P = (A, V, I, C, S, D)$ which quantitatively describes the data forest of predicates. From these, all the measures of Figure 9.13 can be derived with the exception of ΔL. Figure 9.14 summarizes the definition. In the case of a data forest, where $A \le 1$, equations 9.7a and 9.8a of Figure 9.14 generalize to equations 9.7 and 9.8:

Ref. no. of equation	equation	comment
9.1	$T = Y + S$	total no. of term nodes
9.2	$Y = C + X$	total no. of leaves
9.3	$X = V + I$	total no. of variable occurrences where I is the 'identity relation'
9.4	$T = C + V + I + S$	from eq 9.1, 9.2, 9.3
9.5	$L = T$	no. of argument nodes = no. of term nodes
9.6	$L = A + \sum_{i=1}^{i=s} E_i = 1 + \sum_{i=1}^{i=s} E_i$	no. of argument nodes = no. of root nodes (A = 1) + no. of structure-generated argument nodes
9.6a	$L = 1 + 2S_{bin}$	if all structures are binary
9.7a	$1 + S_{bin} = V + I + C$	from equations 9.4, 9.5, 9.6a for binary structures
9.8a	$1 + S + \Delta L = V + I + C$	Δ L is introduced as a correction factor for the general case, where not all structures are binary

Figure 9.13: The inter-relationships of data parameters of a data tree

$$A + S_{bin} = V + I + C \dotfill \text{Eq 9.7}$$

and

$$A + S + \Delta L = V + I + C. \dotfill \text{Eq 9.8}$$

For a given predicate the parameter values may be read off the model, or from the Prolog text directly, by inspection. This 6-tuple is the measure chosen for characterizing all Prolog data objects. Composition of Prolog measures requires operators on this 6-tuple.

parameter name	description
A	the number of roots (and number of trees) of the data forest
V	the number of distinct variables of the data forest
I	the number of binary 'identity relations' creating copies of variables in the data forest
C	the number of constants in the data forest
S	the number of structures in the data forest
D	the maximum depth of structure in the data forest

Figure 9.14: The 6-tuple measure of data forests

The 6-tuple measures of each type of Prolog entity are now developed in turn, starting from elements of simple logic texts (Sections 9.3.3 - 9.3.8) and continuing to texts which include the residual features (Section 9.3.9). The measurement process is given in the summary Section 9.5.

9.3.3 Measures of the body of a clause

We adopt the convention of designating the 6-tuple and its parameters by a suffix to refer to the predicate which is being measured. The suffix 'b' is reserved for the body of clauses. The clause body is a set of predicates. As such, its 6-tuple $P_b = (A_b, V_b, I_b, C_b, S_b, D_b)$ can be read off directly from its model, or from the Prolog text. However, the measure may also be obtained indirectly, by mapping the conjunctive composition of the clause body's constituent predicates into the measurement domain. The first composition operation on the measures is thus one of conjunctive addition.

Conjunctive addition operator

The conjunctive composition '•' over 6-tuples is defined such that :

$$\bullet : N^6 \times N^6 \to N^6$$

$$(A_1, V_1, I_1, C_1, S_1, D_1) \bullet (A_1, V_2, I_2, C_2, S_2, D_2) = (A, V, I, C, S, D)$$

where $A = A_1 + A_2,\ V = V_1 + V_2,\ I = I_1 + I_2,\ C = C_1 + C_2,\ S = S_1 + S_2,$
and $D = \text{max\_of}(D_1, D_2)$.

Both + and max_of are associative and commutative, therefore the binary operator • is also associative and commutative. Hence the operator is n-ary and may be used in the form:

$$\bullet(\text{6-tuple}_1, \ldots, \text{6-tuple}_n).$$

The identity is (0,0,0,0,0,0). Since max_of has no inverse, the conjunctive addition operator has no inverse.

The equality $A + S = V + I + C$ for binary structures is preserved over conjunctive composition if it holds over each of the operands.

Using this definition, the 6-tuple measure of the clause body is obtained by first measuring each constituent predicate, then conjoining their measures. The measure must be adjusted for inclusion in a common scope, decreasing the number of variables V by the number of new identities formed, and increasing I by the same number. We do this by conjoining a modifying 6-tuple. In practice, the majority of identities are created across predicates so that I may be regarded as the number of identities created by the common scope. The scope modification 6-tuple, P_{scope}, is thus $(0,-I, I, 0, 0, 0)$, so that:

$$P_b = \bullet\ (P_{pred1}, P_{pred2}, ..., P_{predn}, P_{scope}).$$

9.3.4 Measures of the head of a clause

The head is a single predicate whose isolated measure is

$$P_h = (\ A_h, V_h, I_h, C_h, S_h, D_h).$$

This is its measure as a fact clause. However, in the context of a rule clause, the head measure is modified as shown in the next section.

9.3.5 Measures of the clause

The clausal model is not a data forest. The parameter A in the procedure and program now models the number of maximal subtrees in the digraph (Fig.9.11).

In defining a clausal implication operator we take note of principle (vi) of the introduction to this section which requires that measures shall be invariant over the internal structure of the Prolog code. In unfolding, body predicates are replaced by the bodies of matching procedures. In this replacement, the heads of the procedures and of their constituent clauses are lost, and hence it is inappropriate to incorporate the number of arguments and number of variables in the head of a clause in the clausal measures. However, due to unification, the number of identities, constants and structures is retained.

Clausal implication addition operator

We define the clausal implication addition operator:
$$\leftarrow\ :\ N^6 \times N^6 \to N^6$$
$$(A_h, V_h, I_h, C_h, S_h, D_h) \leftarrow (A_b, V_b, I_b, C_b, S_b, D_b) = (A, V, I, C, S, D)$$
$$\text{where}\quad A = A_b,\ V = V_b,\ I = I_h + I_b,\ C = C_h + C_b,\ S = S_h + S_b,$$
$$\text{and}\quad D = \text{max\_of}(D_h, D_b).$$

Inspection of the clausal implication addition operator shows that it is equivalent to conjunctive addition of the body 6-tuple measure with a modifying 6-tuple, $(0,0,I_h,C_h,S_h,D_h)$, designated $P_e(head)$. There is no scope adjustment in the single head predicate.

$$P(clause) = \leftarrow (P(head), P(body))$$
$$= \bullet (P_e(head), P(body)).$$

9.3.6 Measures of a procedure

Procedures (matching disjoint clauses) are modelled as a supergraph of the digraphs of individual clauses. The measures mirror the model.

The disjunctive expansion operator

We denote the disjunctive expansion operator by the symbol $|$ and define it such that:

$$| : N^6 \times N^6 \rightarrow N^6$$
$$(A_1, V_1, I_1, C_1, S_1, D_1) \mid (A_2, V_2, I_2, C_2, S_2, D_2) = (A, V, I, C, S, D)$$
$$\text{where} \quad A = max\_of(A_1, A_2), \quad V = max\_of(V_1, V_2), \quad I = max\_of(I_1, I_2)$$
$$C = max\_of(C_1, C_2), \quad S = max\_of(S_1, S_2)$$
$$\text{and} \quad D = max\_of(D_1, D_2).$$

'max_of' is commutative and associative, therefore $|$ is commutative and associative. The operator is n-ary and may be used in the form

$$| (\text{6-tuple}_1, \ldots, \text{6-tuple}_n).$$

The identity is $(0,0,0,0,0,0)$. There is no inverse.

The equality $A + S_{bin} = V + I + C$ is lost over disjunctive composition.

9.3.7 Measures of programs

Measures obtained over fully resolvable hierarchies are given the symbol M. If a hierarchy is fully resolvable to fact-clauses then together with the start predicate it is, by definition, a program. Since the data of the start predicate is not often defined in program text, it is not included in the program measure.

The generic form of the measure of a program is

$$M(prog) = (A, V, I, C, S, D).$$

In Figure 9.15 we identify the elements of the 6-tuple of the program measure with meaningful features of the hierarchy digraph, as shown in Figures 9.10c and 9.11.

The hierarchical substitution operator

This operator governs the measure allocated to predicates in the body of a clause, briefly referred to in Section 9.3.3. The measure for such a predicate is the measure of the matching procedure. The measure of the procedure is substituted as the measure of the body predicate. The hierarchical composition operator is thus one of *substitution*. The value of a body predicate obtained by hierarchical composition differs from that obtained by local inspection of the predicate. It contains measures for all the dependent procedures which make it a program in itself. In contrast to the local P measures, these hierarchically composed measures are designated by M, such that:

$$M(\text{body predicate}) = M(\text{matching procedure}).$$

The M measure for a hierarchical data digraph is calculated recursively by building up from the fact-clausal subtrees, those of the lowest level, following exactly the hierarchical implication process of Prolog shown in Figures 7.5a to c and implemented by the operators discussed above. The measures are easily extracted from the Prolog text directly.

Where the measures of each predicate are not obtained locally but are generated at a lower level and obtained through hierarchical implication, the modifying 6-tuple P_{scope} will be enlarged and will include all locally present constants and structures. We designate such a local effect on the measures as a 6-tuple, $P_e(\text{body}) = (0,-I,I,C,S,D)$.

If the hierarchically obtained measure of a clause body is designated M(body), then such a measure is obtained by conjunctive addition on the 6-tuples:

$$M(\text{body})= \bullet \ (M(\text{pred}_1), M(\text{pred}_2),\ldots,M(\text{pred}_n),P_e(\text{body}))$$

where M(pred) is the measure of each predicate obtained from a lower level.

Remembering that the clausal measure is obtained through the clausal implication operator $\leftarrow$,

$$P(\text{rule-clause}) = \leftarrow(P(\text{head}), P(\text{body}))$$
$$= \bullet \ (P_e(\text{head}), P(\text{body})).$$

The hierarchically composed measures become:

$$M(\text{rule-clause}) = \leftarrow(P(\text{head}), M(\text{body}))$$
$$= \bullet \ (P_e(\text{head}), M(\text{body})),$$

and substituting the expression for M(body) shown above, we get:

$$M(\text{rule-clause}) = \bullet \ (P_e(\text{head}),M(\text{pred}_1, M(\text{pred}_2)\ldots M(\text{pred}_n),$$
$$P_e(\text{body})).$$

parameter name	description and link with the model.
A	the number of arguments which are roots of maximal subtrees (see figure 9.11 and associated text)
V	the number of distinct variables of the hierarchical data digraph calculated as those of the leaves of the maximal subtrees, reduced by any identities in the digraph(this excludes variables in isolated nodes)
I	the number of binary 'identity relations' creating copies of variables in the hierarchical data digraph
C	the number of constants in the hierarchical data digraph
S	the number of structures in the hierarchical data digraph
D	the maximum depth of structure in the hierarchical data digraph

Figure 9.15: Countable features of the hierarchical data digraph

Defining P_e as $\bullet$ (P_e(head), P_e(body)), the clausal measure of a hierarchically composed program is therefore:

$$M(\text{rule-clause}) = \bullet(P_e, M(\text{pred1}), M(\text{pred2})\dots M(\text{predn}))$$

where $\quad P_e = (0, -I_b, (I_h+I_b), (C_h+C_b), (S_h+S_b), \text{max\_of}(D_h, D_b)).$

This expression simplifies calculation of hierarchically composed measures.

Figure 9.15 shows that only the fact-clauses contribute to A and V, and that at any other level contributions (by rule-clauses) are made to C, I, S and D. This is consistent with the requirement that unfolded text returns the same *data* measures as a hierarchically structured text.

The hierarchical measures are calculated as follows:

1. At the lowest level i.e for a single fact-clause,

   $$M(\text{fact-clause}) = P_h(\text{fact-clause}).$$

2. For a body predicate in hierarchical implication by substituting

   $$M(\text{body-predicate}) = M(\text{matching procedure}).$$

3. For a rule-clause by the combined expression for conjunctive and clausal implicative addition over the body predicate measures

   $$M(\text{rule-clause}) = \bullet (P_e(\text{rule-clause}), M(\text{Pred}_1), \dots, M(\text{Pred}_n))$$

   where $\quad$ M(Pred) is the hierarchically composed measure of a body predicate.

4. For a procedure of m fact-clauses and n rule-clauses by disjunctive expansion

$$M(\text{procedure}) = \mid (M(\text{fact-clause})_1, ..., M(\text{fact-clause})_m,$$
$$M(\text{rule-clause }_1) , ..., M(\text{rule-clause}_n).$$

M(program) is then the M-measure of the top level procedure. Note that since a fact-clause has no body, its body measure is $P_b = (0,0,0,0,0,0)$.

9.3.8 Measures of hierarchically composed program segments

Program measures (M) are defined in terms of procedures resolvable to fact-clauses. Direct measures (P) relate to single clauses, either fact-clauses or rule-clauses, or to procedures and are purely local measures.

It may be necessary to measure a hierarchical implication where segments are not resolvable to fact-clauses. We make use of the fact that any rule-clause may be transformed into a program by the inclusion of a fact-clause for each of its body predicates. If each of such fact-clauses is the corresponding 'identity' fact-clause, then the consequent program is called a 'partial program'. The measure for an identity fact-clause is denoted P_i and its value for a fact-clause of arity n is

$$M(\text{name/n}) = P_i \,(\text{name}) = (n, n, 0, 0, 0, 0).$$

A SIMPLE EXAMPLE

Consider a the single-clause procedure (not a program):

```
is_type([Name,Address],member):-
            passed_test(Name,yes),
            lives_at(Name,Address).
```

Adding two identity fact clauses passed_test(_, _). and lives_at(_, _). converts the procedure to a program, albeit not a very useful one, and allows M(is_type) of this partial program to be calculated:

$$M(\text{is\_type}) = \bullet \ (P_e(\text{is\_type}),M(\text{passed\_test}),M(\text{lives\_at}))$$
$$= \bullet \ (P_e(\text{is\_type}),P_i(\text{passed\_test}), P_i(\text{lives\_at}))$$
$$= \bullet \ (0, -1, 1, 2, 1, 1), (2, 2, 0, 0, 0, 0), (2, 2, 0, 0, 0,0)$$
$$= (4, 3, 1, 2, 1, 1).$$

END OF SIMPLE EXAMPLE

The replacement of hierarchical programs by identity fact-clauses corresponds to the pruning of the hierarchy tree (Figure 9.16), and allows the effects of selected branches to be removed from the measures. This is a useful device which is exploited in the following section.

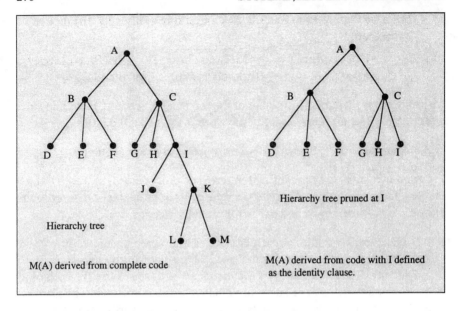

Figure 9.16: Pruning a tree in the Prolog hierarchy

9.3.9 Measures of Prolog texts containing residual features

Measures of directly recursive clauses

Consider a recursive clause such as

```
ancestor(A,B):-
    parent(C,B),
    ancestor(A,C).
```

In modelling recursion the recursive predicate model is replaced by the identity fact clause model, so, in the measures, the recursive predicate measure is that of the corresponding identity fact clause and M(recursive predicate) = P_i(recursive predicate). The measure for the 'ancestor/2 clause above would thus be calculated as

M(ancestor)= •(P_e(ancestor),M(parent), P_i(ancestor)).

Measures of indirect recursion

As in the case of simple recursion, indirect recursion gives rise to body predicates to which we cannot allocate a measure. In the notional code below, we have no measure for do1 until we measure do3, and no measure for do3 until we measure do1.

```
do1:- do2, do3.
do3:- do4, do1.
```

The measure for such a group follows from the model, and takes account of the requirement on a measurement scheme that it should return a unique value. The hierarchically composed measure M(do1) must equal the hierarchically composed measure M(do3), since they differ only in the entry point to the same indirectly recursive group, and the composite measure must be symmetrical with respect to do1 and do3.

This requirement is met by the measures of the group model given previously, in section 9.2.3 and where the identity measure P_i is used for each indirectly recursive predicate:

$$M(clause_1, \ldots, clause_n) = \bullet(P_e(clause_1), \ldots, P_e(clause_n), P_i(clause_1),$$
$$\ldots, P_i(clause_n), M(predicate_1), \ldots, M(predicate_m)).$$

where the clauses are those in the recursive cycle, and the M measures are those for any other body predicates not in the cycle. For the notional code above, the measure would be expressed as:

$$M(do1, do3) = \bullet(P_e(do1), P_e(do3), P_i(do1), P_i(do3), M(do2), M(do4).$$

With only one clause in the cycle, i.e. for direct recursion, the measure degenerates to that proffered for directly recursive clauses.

Measures of system predicates

System predicates were discussed in Section 7.3.3. Some, such as read/1 and write/1, serve input/output communication and other operational requirements; these are not part of the logic of a Prolog program and are to be removed before measurement, as most other system predicates associated with system handling, garbage collection and debugging. Mathematical operations such as addition and subtraction are also excluded. Some predicates are, however, intrinsic parts of the logic. Subjective decisions will have to be made by the designer of the measurement scheme as to what to include in the static measurement. We have found it necessary to include the comparison operators, $>$, $<$, $\leq$, $=$, $==$, etc. Excluding metalogical predicates, which are discussed below, for any retained system predicate of arity n, its measure is $(n, n, 0, 0, 0, 0)$, the identity measure.

Measures of metalogical predicates

The 'metamodel' illustrated in Figure 9.9 gives a clear picture of how the measurement of the metalogical clauses may logically be fitted into the measurement scheme, although we can no longer refer to the contents of arguments of such clauses as data objects. We shall take each 'argument +

content' of a metalogical predicate and measure it as a separate data forest. For
the model of Figure 9.9, which represents the code:

```
intersect(Xs,Ys,Zs):-
    setof(X,(member(X,Xs),member(Y,Ys)),Zs).
```

The 'arguments + contents' of the clause body yield the measures (1,1,0,0,0,0),
M(member/2)•M(member/2) and (1,1,0,0,0,0) respectively. These four conjoined with P_e(Intersect/3) give the measure for the definition 'intersect/3'.

In general, for any metalogical clause, the clause measure is given as the
conjunctive composition of P_e for the clause and the measure generated by
each of the 'arguments + contents' measured as individual programs.

9.4 An example of measurement

We now return to the EXAMPLE PROGRAM 'union/3' and its subprogram
'member/2', given in Section 9.2.2. The measures are first obtained by hand
from the text: each clause of the program is given separately below, followed
by the derivation of its measure and the measure calculated 'bottom-up' using
the definitions of the measures in Section 9.3. This example is simple enough
for the measures to be obtained by inspection of the model, but for larger Prolog
texts the measures need the assistance of an automated measurement tool.
PDQ was produced to assist the measurer [2]. The hand-worked measures of
the example are followed by the PDQ output for the same program, and are
shown in Figure 9.17. The measures obtained by inspection of the model agree
with the hand-calculated measures and the PDQ measures.

EXAMPLE PROGRAM 'member/2'

1a member(H,[H|Tail]).
　Fact-clause
　M(member/2-clause 1) = P_h(member/2 clause 1)
　　　　　　　　　　　 = (2, 2, 1, 0, 1, 1)

1b member(Element,[H|Tail]):- member(Element,Tail).
　Recursive rule-clause
　M(member/2-clause 2) = P_e(member/2-clause2) • M(identity/2)　　　*conjunction*
　P_h(member/2-clause2) = (2, 3, 0, 0, 1, 1), P_b(member/2-clause2) = (2, 2, 0, 0, 0, 0)
　∴　P_e(member/2-clause2) = (0, 0, 0, 0, 0, 0, 1, 1)
　∴　M(member/2-clause 2) = (0, 0, 0, 0, 0, 0, 1, 1) • (2, 2, 0, 0, 0, 0 = (2, 2, 0, 0, 1, 1)
　∴　M(member/2)　= M(member/2-clause 1) I M(member/2-clause 2)　*disjunction*
　　　　　　　　　　 = (2, 2, 1, 0, 1, 1) I (2, 2, 0, 0, 1, 1 = (2, 2, 1, 0, 1, 1)

END OF EXAMPLE PROGRAM 'member/2'

EXAMPLE PROGRAM 'union/3'

2a. union([],Set,Set)*.
Fact-clause

M(union/3-clause 1) = P_h(union/3 clause 1) = (3, 1, 1, 1 ,0, 0)

2b. union([Element|Rest],Set,Union):- member(Element,Set),!, union(Rest,Set,Union).
Recursive rule-clause

P_h = (3, 4, 0, 0, 1, 1) P_b= (5, 4, 1, 0, 0, 0, 0)

∴ P_e = (0, -1, 1, 0, 1, 1)

M(union/3-clause2) = Pe(union/3-clause2) • M(member/2) • M(identity/3)
                    = (0, -1, 1, 0, 1, 1) • (2, 2, 1, 0, 1, 1) • (3, 3, 0, 0, 0, 0)
                    = (5, 4, 2, 0, 2, 1)

2c. union([Element|Rest],Set,[Element|RestUnion]:- union(Rest,Set,RestUnion).
Recursive rule-clause
P_h = (3, 4, 1, 0, 2, 1) P_b= (3, 4, 0, 0, 0, 0)

∴ P_e = (0, 0, 1, 0, 2, 1)

M(union/3-clause3) = Pe(union/3-clause3) • M(identity/3)
                    = (0, 0, 1, 0, 2, 1) • (3, 3, 0, 0, 0, 0) = (3, 3, 1, 0, 2, 1)
∴ M(union/3) = M(union/3-clause 1) | M(union/3-clause 2) | M(union/3-clause 3)
              = (3, 1, 1, 1 ,0, 0)| (5, 4, 2, 0, 2, 1)| (3, 3, 1, 0, 2, 1) *disjunction*
              = (5, 4, 2, 1, 2, 1)

* An empty list is a functor with no arguments and thus counts as a constant, see 7.4..2

END OF EXAMPLE PROGRAM 'union/3'

Clause	Ph	Pb	Pe	M1	M
member/2/1	2,2,1,0,1,1	0,0,0,0,0,0	2,2,1,0,1,1	2,2,1,0,1,1	
member/2/2	2,3,0,0,1,1	2,2,0,0,0,0	0,0,0,0,1,1	2,2,0,0,1,1	2,2,1,0,1,1
union/3/1	3,1,1,1,0,0	0,0,0,0,0,0	3,1,1,1,0,0	3,1,1,1,0,0	
union/3/2	3,4,0,0,1,1	5,4,1,0,0,0	0,-1,1,0,1,1	5,4,2,0,2,1	
union/3/3	3,4,1,0,2,1	3,3,0,0,0,0	0, 0,1,0,2,1	3,3,1,0,2,1	5,4,2,1,2,1

Data measurements for program union

Figure 9.17: PDQ output on program 'union'

9.5 The measurement scheme – a summary

The measurement scheme is composed in accord with the general principles of model-based measurement. The models retain features of the Prolog text selected in Chapter 7, and the measures reveal the inherent properties of the data model, in accord with the principles set out in Section 9.3.

9.5.1 Defining measures for the 'foundation level' of the data model

The models developed in Section 9.2 expose the data as a structure over the lowest-level elements to which the text resolves. This level forms the foundation on which the data of the Prolog logic text is built. The structure of the higher levels is also of interest, but it does not alter the content of the data. The parameters listed below include both explicit features and those implicit in the foundation level.

- Parameter A designates the number of predicate *argument nodes* in the bottom data layer. This parameter gives the number of arguments in the fact-clauses to which the data entity resolves.

- Parameter V is the cardinality of the *set of variables* V. The *cardinality of the bag* is given in the model as the number of simple term nodes at the bottom of the digraph. To calculate the *cardinality of the set*, we subtract from this the number of 'identities' (denoted by the equality sign) in the model, whose effect is transmitted to the lower levels.

  This measures the number of dimensions of the state space of the data entity, a property for characterizing the 'size' of the problem addressed.

- Parameter I is the *number of identities*. It represents data utilization within the data entity: the extent to which it makes use of the incoming data *.

- Parameter C is the *number of constants* *. It refers to the information base on which the data entity draws.

- Parameter S is the *number of structures* given by the number of structure bars in the model. It represents the extent to which the incoming data is manipulated inside the data entity *.

- Parameter D is the *maximum depth of structure* *.It indicates the maximum degree of nestedness in the data organization.

  * Values at all levels will be passed to the foundation level by unification and are therefore implicit in that level.

These six parameters are collected together as a 6-tuple (A,V, I, C, S, D).

9.5.2 Defining measures for other data entities

Using the concept of structural models, the composition of larger data entities are composed from their constituents, as described in Section 9.2.2. Mirroring this, measures can be composed for the whole program and all intermediate data entities of the Prolog logic text by applying composition operators to the 6-tuples of the foundation-level data entities.

- A conjunctive composition operator involves summation for each of the first 5 parameters and taking the maximum of the last parameter and is defined as:

$$\bullet : N^6 \times N^6 \to N^6$$

$$(A_1, V_1, I_1, C_1, S_1, D_1) \bullet (A_2, V_2, I_2, C_2, S_2, D_2) = (A, V, I, C, S, D)$$

where $A = A_1 + A_2$, $V = V_1 + V_2$, $I = I_1 + I_2$, $C = C_1 + C_2$,
$S = S_1 + S_2$, and $D = \text{max\_of}(D_1, D_2)$.

- A clausal implication operator is

$$\leftarrow : N^6 \times N^6 \to N^6$$

$$(A_h, V_h, I_h, C_h, S_h, D_h) \leftarrow (A_b, V_b, I_b, C_b, S_b, D_b) = (A, V, I, C, S, D)$$

where $A = A_b$, $V = V_b$, $I = I_h + I_b$, $C = C_h + C_b$, $S = S_h + S_b$,
and $D = \text{max\_of}(D_h, D_b)$.

- A disjunctive composition operator derives the maximum of each of the parameters:

$$| : N^6 \times N^6 \to N^6$$

$$(A_1, V_1, I_1, C_1, S_1, D_1) | (A_2, V_2, I_2, C_2, S_2, D_2) = (A, V, I, C, S, D)$$

where $A = \text{max\_of}(A_1, A_2)$, $V = \text{max\_of}(V_1, V_2)$, $I = \text{max\_of}(I_1, I_2)$
$C = \text{max\_of}(C_1, C_2)$, $S = \text{max\_of}(S_1, S_2)$
and $D = \text{max\_of}(D_1, D_2)$.

- An hierarchical composition operator is merely one of substitution. The procedure measures of one level are substituted as the measures of the matching body predicates of the next higher level.

The mapping of the four operators from Prolog code, through models to measures, is summarized in Figure 9.18.

The composition operators are applied as follows:

- Body measures are obtained by conjunctively composing lower level measures for each body predicate with a modifying measure P_e(body) which allows for common scope and local constants and structures.

- Clause measures are obtained by applying the clausal implication operator to the head and the body measures where the head measure is the modifier P_e(head). Since the two modifiers, P_e(head) and P_e(body), may be conjunctively composed to a single clausal modifier P_e, the clause measure is simply obtained by conjunctively composing the measures of the body predicates with P_e.

- Procedure measures are obtained by disjunctively composing the clausal measures of all the clauses making up the procedure.

Operators	Prolog	Model	Measure
1	conjunction	conjunctive composition	conjunctive addition on 6-tuples and adjustment for scope: •(M1,M2,...Pe(body))
2	clausal implication	composition over clausal implication	clausal addition adjustment for head of clause •(M(body),Pe(head))
3	disjunction	disjunctive composition	disjunctive expansion M(procedure) =l(clause1,...clausen)
4	hierarchical composition	hierarchical composition	substitution M(Name)=Mproc(Name)

Figure 9.18: Operators

Program measures are derived by the following recursive procedure:

1. Identify the measures of the lowest level fact clauses.

2. Until the highest level procedure is reached:

   2.1 disjunctively ·compose the clausal measures to give procedure measures;

   2.2 assign these procedure measures to the matching body predicates of the higher level clauses;

   2.3 calculate the measures of the next higher level clauses by conjunctively composing the body predicated measures;

   2.4 conjunctively compose an adjustment 6-tuple which includes the effect of common scope and the identities, structures and constants in both head and body predicates.

The measure of the start predicate is omitted by convention since its data objects are either not known or not fixed.

There are three further Prolog features, collectively referred to as 'residual' features and whose models were shown in Section 9.2.3. Their measurement is outside the normal procedure. The chosen models demonstrate how they are to be measured:

- The measure for a recursive predicate is the measure of the corresponding identity predicate, P_i.

- The measure for a system predicate is the measure of the corresponding identity predicate P_i.

- The measure for a metalogical predicate is the conjunctive addition of the measures of each of its arguments taken as independent programs together with the local P_e measure.

9.5.3 Obtaining the measures in practice

Measures can be obtained by using the general procedure of model-based measurement, progressing first from referent (the Prolog logic text) to model, and then reading the parameter values from the model. The process of drawing up the model is tedious: once the measures have been defined through modelling, the measurement process can bypass the model. The values of measures for foundation level entities can be obtained directly from the Prolog logic text, and measures for the rest of the code can be derived indirectly from them, by composition.

This process works well for small Prolog texts, but for sizeable programs the reading and hand-crafting of measures is lengthy and error-prone. This process is easily automated, and just such a tool (Prolog Data Quantifier or PDQ) has been produced and is available. The tool will accept any syntactically correct and complete Prolog code, generate from it the Prolog logic text, and returns the composed 6-tuple measure for the whole program in detail, or for any of its parts.

9.5.4 Constructing the measurement scheme

We refer to the general model-based measurement scheme of Figure 3.8.

The individual parameters of the 6-tuples of foundation-level data entities are the (directly or indirectly measured) property variable measures at the bottom layers of the measurement scheme. For the program, or for any of its intermediate components which is of interest as a referent in its own right, the complete 6-tuple is the object-oriented measure of the scheme. Any feature, such as those of Figure 9.13 not in the 6-tuple, may be derived from these six parameters

In Section 9.5.1 we discussed the interpretation of the individual parameters of the 6-tuple of the foundation-level data entities. This interpretation applies for all levels of data entities, including the program as a whole.

The construction of a 'complexity measure', or some other utility measures for the Prolog logic text, is beyond the scope of this book, but we offer some speculative ideas.

Some possible utility measures

- Each program parameter of the 6-tuple represents some aspect of the intuitive notion of 'complexity'. Thus, the larger the parameters, in aggregate, the more complicated the referent data entity. Bearing this monotonicity in mind, the sum of the six parameters would be a simplistic 'complexity measure' of the data in the Prolog logic program.

- Recall the relationship between *arguments* and *terms* (equation 9.8), where for a data forest

$$A + S + \Delta L = V + I + C.$$

and $\Delta L = 0$ for only binary structures.

The values aggregated over a program are based on the data forests at the lowest level of the hierarchy digraph (Figure 9.15), corrected for identities and structures at higher levels. Hence we may use either side of the equation to give a 'size' measure in terms of the number of arguments (or equivalently the number of atomic data terms) in the program. The model enables us to interpret this as a maximum size over the disjuncts. In practice it is more convenient to use $V + I + C$ as the size measure, since this avoids the problem of ΔL.

- Given a size measure, the extensive (additive) parameters I and S may be used to derive the 'intensive ratios'

$$\frac{I}{(V+I+C)}, \quad \frac{S}{(V+I+C)} \quad \text{and} \quad \frac{I+S}{(V+I+C)}.$$

In a simple Prolog clause the main method of imposing meaning is by assigning identities and structures. It then follows that $(I + S)$ measures the 'manipulation' of data in a program, and reflects on its 'difficulty'. In an experiment carried out during research work for British Telecom ([3]), the ratios given above showed remarkable consistency within individual programs while varying between programs, but to date only a small set has been examined.

- The model of Prolog data, as currently proposed, preserves many of the interesting attributes of the logic text, but, as all models, it is selective. For example, the model does not preserve the functors of the program text. The parameters represent the cardinality of the *bag* of functors, but do not allow the measurement of the size of the *set*, which would indicate variety, a feature of complexity. If this parameter is of importance in a given application, it can be provided by a slight modification of the model (labelling the structure nodes), and a corresponding expansion of the measures.

- It is an important property of the measurement scheme that the same mathematical structure (a six-tuple of integers) characterizes the whole program, and any of its parts, right down to the sub-clausal entity of a single predicate. This 'group property' of the measures facilitates the quantitative appraisal of the design from the viewpoint of complexity management. Hierarchical design incorporates intermediate structures which are logically redundant but contain the explosion of detail. In an ideal case each intermediate procedure, and each clause, would need to justify its existence by carrying some of the burden of expanding complexity. The *distribution* of measures among the program's parts would show the extent to which the design is 'fair' in equitably carrying the burden.

A concluding comment

The 6-tuple, as presently devised, is once again representative rather than definitive. It captures only some selected items of information present in the model. For example, the model would permit counting the *maximum dispersion* of data structures (maximum of the out-degree of structure nodes), or the *average out-degree*. It was our choice not to include these in the measure set, but the set can be re-designed if required, to include some other attributes preserved in the model.

The use of the measurement scheme is not restricted to the assessment of Prolog program text. In cases where the information in the program is of interest but the overt structure of the program is not, such as in declarative specification, aggregated data measures offer grounds of estimating the size and logical content of the information stored in the specification as proposed in Chapter 6 and illustrated in Chapter 10.

We remind the reader of the generic measure for product specification, developed in Section 5.2.4 and 5.2.5, which was of the form:

$$M_{obj}(S) = (p, r, z, m, y, t),$$

where the property measures are p, m and z.

These are fundamental for system specification, and are related to the group measures of the data of Prolog text. Looking at the data measures of the 6-tuple (A, V, I, C, S, D), we note that $V + C$ represents the cardinality of the property set. A component of p, $C / (V + C)$, corresponds to z. S is the cardinality of the set of relations over the properties, and is thus a component of r. D is a tree measure on the relations over the properties and is thus part of y. Only the time component is missing. The exception is I, the 6-tuple parameter signifying the degree to which it is necessary to repeat properties in order to express the ideas in the specification. This parameter is not present in the general measure.

The general measures and the specific Prolog data measures were developed independently. The reliance of the general measure on systems principles, and the substantial agreement between the two measures, gives added confidence in the soundness of the Prolog data measures. The implication of the difference with respect to the parameter I is to be explored in further work, when the Prolog data measurement scheme is next revised. This demonstrates the way in which model-based measurement progresses.

9.6 References

1   Myers M, Kaposi A A (1991): "Modelling and measurement of Prolog data". Software Engineering Journal, Vol 6, No 6, pp 413-434. IEE.

2   PDQ User Manual in: Myers M (1992): "Model-based measurement of formal and semi-formal specifications". Kaposi Associates Internal Report G23.

3   Myers M (1990): "Evaluating Specifications and Designs – Worked examples on the demonstrator PDQ". Contract ML316214, RT31, BT, Martlesham.

10 AN EXPERIMENT IN MEASURING SPECIFICATIONS

10.1 Introduction

The aim of this chapter is to demonstrate the multiple-language strategy of model-based measurement outlined in Section 6.5. The material presented here is drawn from a comprehensive report ([1]), and sets out the assumptions and procedures of the case study. Of necessity, the treatment here is much curtailed. The aim of the experiment itself is to show how to carry out model-based measurement of formal and semi-formal specifications. The measurement proceeds in accord with the multiple-language strategy described in Figure 6.10.

The **apparatus** of the experiment comprises the collection of *languages* in which the specifications are expressed, the collection of *translators* which yield all the specifications in the High Level version of the reference language Prolog, the *Mathematical Toolkit* which implements the specifications in Full Prolog, and the *measurement tool PDQ*, which automatically produces the measures from the Full Prolog. The apparatus, assembled from research by the authors and others, is described in Section 10.2.

The **method** of the experiment implements the multiple-language strategy in five stages, outlined in Section 10.3, and the **results** of each stage are summarized in Section 10.4, with the aid of the Appendix.

Section 10.5 gives the **conclusions** of the experiment, including a brief discussion of the measures.

A full description of the experiment explains the use of the specification languages through the chosen example. It shows how the measures are composed, bottom-up, from measures of the smallest meaningful element of the specification to the measures of the specification as a whole. The full report also contains further examples to allow more informed general discussion of the factors which influence measures in the measurement scheme.

10.2 Apparatus

10.2.1 Specification languages

Four specification media were chosen, Z, VDM, Data Flow Diagrams and State Transition Formalism. In addition, Prolog itself – the reference language of the multiple-language strategy – is also used as a specification language.

Z

Z is a model-based specification language based on First Order Predicate Logic and typed set theory. It is now sufficiently well known through a number of textbooks. Its notation was first defined in [2]. The presentation of Z has varied slightly in subsequent publications, but recent work on a formal syntax indicates a consensus of those in the field.

In Z, specifications are represented by 'schemas', produced in either vertical or horizontal form (as in Figures 10.1 and 10.2). The first part of a Z schema contains the type declarations. The second part contains the axioms.

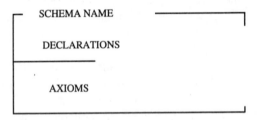

Figure 10.1: Vertical form of Z Schemas

Schema ≜ DECLARATIONS | AXIOMS

Figure 10.2: Horizontal form of Z Schemas

VDM

VDM (the Vienna Development Method) employs a specification language developed at IBM Vienna laboratories [3]. The language, too, is known as VDM. Like Z, VDM is model-based, and uses logic and mathematical notations. VDM specifications are presented as a set of types, pre-conditions, post-conditions and invariants. The specifier is obliged to prove that the post-conditions are achievable from the pre-conditions by using a function which connects the state variables.

Data Flow Diagrams (DFD)

There have been several methods of 'structured systems analysis' developed over the past two decades. These contain both prescriptive methods and graph models, see e.g. (4). A model common to these methods is the Data Flow Diagram (DFD) and its accompanying Data Dictionary. The representation is freer than that of the other methods introduced here, which makes for easier accessibility for the majority, but demands greater care in application. There is no associated reasoning system.

Data Flow Diagrams comprise a graphical specification method. Compilation of these diagrams increases knowledge of, and confidence in, a proposed system by helping system designers and developers understand and check data flow through the system. They have been found particularly helpful in designing systems involving complicated data processing.

There are three types of object in a DFD:

- *stores*, whose symbol is a name positioned between two horizontal lines;
- *processes*, symbolised by named circles,
- *outside entities,* represented as named rectangles;

The three types of object are connected by arrows, signifying data flow.

The contents of the stores show the *state* of the system at any one time, and the processes are the *operations* by which a change of state is brought about. The outside entities are the agents by which processes are initiated and/or recipients of process output.

The Data Dictionary has sections corresponding to stores, data flows and processes. The stores entry will describe all the data objects in the store, and roughly corresponds to type statements in more rigorous languages. The data flow entries detail which data is flowing along the arrows. The entries describe the processes in as much detail as the specifier deems necessary.

State Transition Formalism (STF)

The State Transition specification method, illustrated in Chapter 5, labels states without detailing their black-box specifications. Part 1 of the specification, it will be recalled, is the set of states of each of the two components and their cross-products. Part 2 is the forced state transition table, and Part 3 is the free-will state transition table. Any of the states may then be represented as a list of parameters describing that state.

10.2.2 Translations into Prolog

In all cases, the method of translation is reasonably clear on comparison of the specifications and the relevant resultant code texts. However, the tutorial approach of ([1]) makes each translation step explicit. The following notes direct the reader to the relevant source references.

Z

A project carried out by Knott and Krause ([5]) investigated the translation of Z into Prolog. This was found to be feasible, and the work resulted in their producing a Z procedure library, in which each element of the Z notation is represented by an equivalent Prolog procedure. The purpose of Knott and Krause's translation was animation of Z specifications. To that end, the initial Prolog code produced by a straight one-to-one rewrite was subjected to a series of transformations ([6]). Whilst the characteristics of the transformed code will eventually be of interest for comparison purposes, here we are only concerned with translation.

VDM

A study of the translation of VDM into Prolog is given by White ([7]). Here again, the purpose of investigating the translation was validation of a VDM specification by execution, and Prolog was used both to produce an animated specification and as a prototype medium.

Data Flow Diagrams (DFD)

Goble, in ([8]), shows how DFD's translate into Prolog. As with the other translations exemplified above, the motivation of that work was validation of the specification through execution. His instructions are simple and explicit:

for each process box:

- make the exit point the head of a Prolog rule,
- place the inputs in the body of the rule,
- perform the necessary process in the body and pass the results to the head.

State Transition Formalism (STF)

A desire to animate specifications led to the translation of the State Transition Formalism into Prolog, first as an expression of the transitions as Prolog statements, and then as implementation of a harness to chain them ([9]).

10.2.3 Mathematical Toolkit

The expansion of High Level Prolog into Full Prolog requires Prolog definitions for all symbols of the specification languages used. A useful terminology comes from Spivey's Z reference book ([10]), where the author entitles a chapter 'The Mathematical Toolkit'.

We embrace this terminology to include all the mathematical definitions used in the specification languages of our experiment. The greater part of the Mathematical Toolkit is common to all the chosen formal notations. Any symbol, special to a particular specification language, is added to this common core. The toolkit thus serves the needs of all the specification languages participating in the measurement scheme. Using the notion of Mathematical Toolkit, we can distinguish between High Level Prolog – the line-by-line translation of a specification into Prolog syntax – and the Prolog text needed to define the mathematical tools. The inclusion of the latter in the Prolog text produces Full Prolog. The Prolog definitions of the items of the Mathematical Toolkit which are needed in the examples of this chapter are given in the Appendix as Code text 10A.7. Many of the definitions are taken from ([5]). Some theoretical difficulties are encountered in such definitions, for example in the representation of sets, and these are dealt with both in the reference just cited, and in the tutorial report ([1]) on which this chapter is based.

The Prolog definitions of the Mathematical Toolkit reveal a hierarchy of concepts which is not immediately obvious, and this is reflected in the measures. A single example is given in Figure 10.3, but a more comprehensive example, with deeper hierarchy, is given in ([1]).

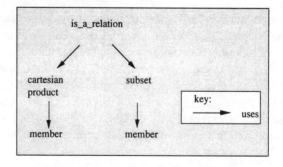

Figure 10.3: Dependence of the 'is_a_relation' definition

10.2.4 PDQ

'Prolog Data Quantifier' (PDQ) is a tool written in LPA MacProlog, produced specifically for automating the measurement of the 6-tuples of Prolog data. It is based on a module of an earlier tool SPA ([11]). PDQ obtains the measures of Prolog entities by following the algebra of Chapter 9, composing recursively from the lowest to the highest levels. Examples of output are given in the appendix to this chapter and the PDQ User Manual is in ([1]).

By making use of predefined libraries, PDQ can be used in two modes. Given a syntactically correct Prolog logic text, one mode produces the High level Prolog measures and the other mode produces the Full Prolog measures.

10.3 Method

The method of measurement is based on the multi-lingual deductive strategy of Chapter 6. The experiment has five stages.

Stage 1 identifies the object: a small natural language specification.

Stage 2 selects five different specification media, Prolog, Z, VDM, Data Flow Diagrams and a State Transition Formalism, and expresses the specification in each of them.

In **stage 3**, all but the first of the stage 2 versions are translated line by line into the common reference medium, High Level Prolog.

These first three stages are illustrated in Figure 10.4.

In **stage 4** the High Level Prolog specifications are expanded to Full Prolog, using the Mathematical Toolkit definitions. The original High Level Prolog specification is declarative, as is the mathematical formulation from which it is derived. If the Mathematical Toolkit is specified in like manner then the Full Prolog version will remain a declarative representation of the original specification. Figure 10.5 shows the role of the Mathematical Toolkit in stage 4 of the experiment. The Toolkit needs to be formulated in Prolog only once; thereafter it is available for general use.

In **stage 5**, all sets of codes so produced, both High Level and Full Prolog as appropriate, are measured, using the PDQ tool and the results recorded. Figure 10.6 shows the measurement stage of the experiment in context.

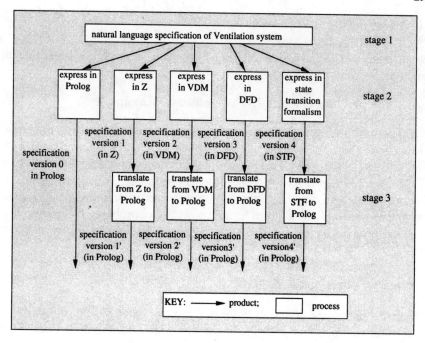

Figure 10.4 Outline of experiment, stages 1, 2 and 3

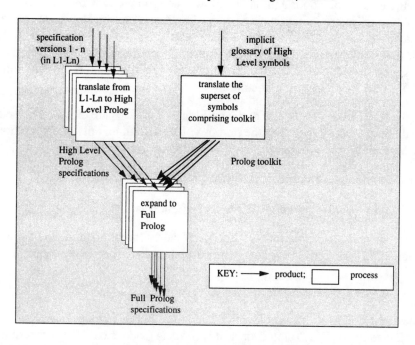

Figure 10.5: Outline of experiment, stage 4 – completion of the text by creation of Full Prolog specifications

10.4 Results

We now discuss the product resulting from each stage of the experimental process.

10.4.1 The product of stage 1: The 'ventilation' example

When demonstrating new methods, the choice of example poses a dilemma. The example must be small and readily understandable, so that the reader's attention can be focused on the method, rather than be distracted by the details of the example itself. At the same time, the example must have sufficient features to demonstrate the method. With these considerations, we extracted a simplified example from an environmental control problem. A different and more comprehensive example, specifying a complete music library, is given in ([12]).

The natural language description

The entity under specification is the rudimentary ventilation system of an industrial building. In order to restrict the size of the specification, the stylized building (MyBuilding) is assumed to consist of a set of rooms, with each room having exactly one door and one window. Each room opens from a corridor: there are no intercommunicating rooms. The doors and windows are known collectively as 'Orifices'. Orifices may be in one of two states, 'open' or 'closed'. To provide correct environmental conditions for the industrial processes carried out in the building, it is ruled that, in each room, only one orifice may be open at any one time.

At the start of the service life of the building, all orifices are closed. This sets up the initial conditions. For the purpose of the example, only one operation on MyBuilding is considered, that of opening a window.

The 'Ventilation Example' is illustrated in Figure 10.7.

10.4.2 The product of stage 2: The specifications

Other than the designation of the specification medium and the natural language description of the referent, no instructions were given to the specifiers.

Direct Prolog specification

The construction of this specification from the natural language description is explained in ([1]). The Prolog code as it appears in Code text A10.1 was written declaratively, without regard for operational efficiency. An alternative version of the code, Code text A10.2, shows an improvement in this respect.

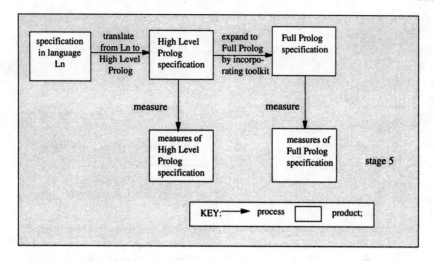

Figure 10.6:
Outline of experiment, stage 5 – the measurement processes for a language Ln

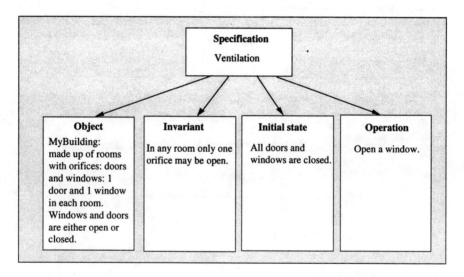

Figure 10.7: Structure of specification for ventilation system

Z specification

The Z schemas (unpublished work by Birznieks) comprising the specification are given without comment. The schemas of Figures 10.8 to 10.13 summarize the specification. The background report ([1]) adopts a tutorial approach, explaining how the specification is constructed.

[ORIFICE]

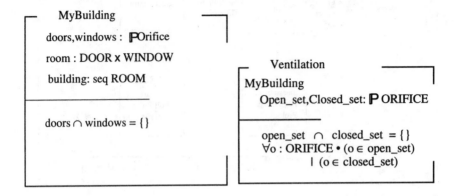

Figure 10.8: The Z schema for MyBuilding   **Figure 10.9**: Ventilation schema

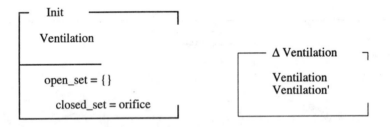

Figure 10.10: The initial state schema   **Figure 10.11:** A delta schema

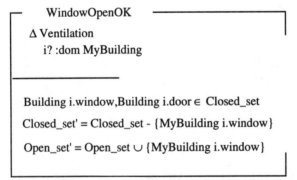

Figure 10.12: Schema for successful opening of window

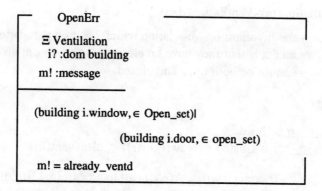

Figure 10.13: Schema for an erroneous request

VDM specification

The VDM Specification (unpublished work by Birznieks) is given below. Some comments are included in the VDM comment format.

<u>State</u>

Building ::	rms : Room-set
	vent : Ventilation
Room ::	w : Orifice
	d : Orifice
Ventilation ::	opn : Orifice_set
	clsd : Orifice_set
Orifice_set = to be defined	

<u>Invariants</u>
inv-Building(mk-Building(rs ,v))
$\Delta \forall r \in rs \bullet \neg (\{w(r),d(r)\} \subseteq Opn(r))$
$\land (\forall s \in rs \bullet \neg (d(r) = w(s))$
$\land (w(s) \neq w(r) \lor r = s)$
$\land (d(s) \neq d(r) \lor r = s))$

--line (1) The invariant on Building with parameters rs and v
--line (2) is defined as: for every room its window and its door cannot both be open, so together they cannot be a subset of the set of open orifices,
--line (3) and no orifice can be both a door and a window
--line (4) and if a window of room r is the same as the window of
--   room s, then room r is the same as room s,
--line (5) and if a door of room r is the same as the door of room s, then
--   room r is the same as room s.

inv-Ventilation(mk-Ventilation(o,c)) $\underline{\Delta}$ o $\cap$ c = { }

> --the invariant on Ventilation which has two parameters, sets o
> --and c, is that they have no elements in common; an orifice
> --cannot be both open and closed.

INIT()

> ext wr b:Building
> post $\forall$ r $\in$ rms(b) • {wdw(r),dr(r)} $\subseteq$ clsd(vent(b))
>
> --There is a w(rite)/r(ead) call to an external Building,
> --followed by a definition of the state of the building on
> --acceptance.
> --There is no need to specify that opn(vnt(b)) = $\varnothing$ as this
> --follows from the invariant inv-Ventilation. These
> --invariants are assumed conjuncts to all operation pre-
> --and post-conditions.

OPENWINDOW_OK (r:Room)

> ext wr b:Building
> pre r $\in$ rms(b) $\wedge$ {wdw(r),dr(r)} $\subseteq$ clsd(vent(b))
>
> post opn(vent(b)) = $\overset{\diagup}{\text{opn(vent(b))}}$ $\cup$ {wdw(r)}
>
> --The hook signifies the previous value of a function or
> --variable.
> --Again invariants demonstrate that wdw(r) does not belong
> -- to the open set of ventilation prior to the operation, and
> --that the altered state of clsd(vent(b)) follows automatically
> --from the post-condition.

ERROROPEN W(r:rms) msg: string

> ext   r b : Building
> pre r $\in$ rms(b) $\wedge \neg$ {wdw(r), dr(r)} $\subseteq$ clsd(vent(b))
> post msg

DFD specification

The Data Flow Diagram specification is given in Figure 10.14, and the three parts of the data dictionary are shown in separate Figures, 10.15, 10.16 and 10.17. Further details are supplied in ([1]).

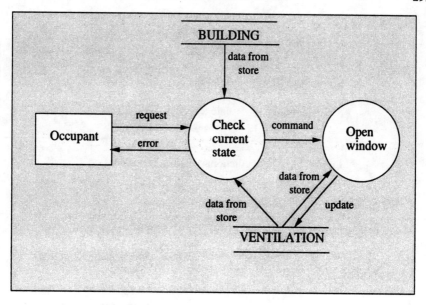

Figure 10.14: Data flow diagram of 'Ventilation'

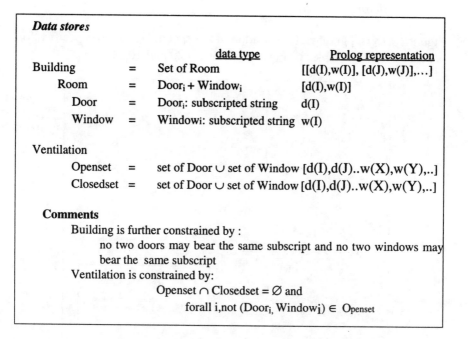

Figure 10.15: Data dictionary Part 1, Data stores

DATA FLOWS

			data type	Prolog representation
request				
	Room identifier	=	integer	I
command				
	Room identifier	=	integer	I
error		=	Message(string or audio)	Msg

Figure 10.16: Data dictionary Part 2, Data flows

PROCESSES

check current state	
	does the room exist?
	are both window and door closed?
open_window	
	change state of window

Figure 10.17: Data dictionary Part 3, Processes

STF specification

Figure 10.18 is a graphical illustration of the specification. The specification is presented in the tables of Figures 10.19, 10.20 and 10.21.

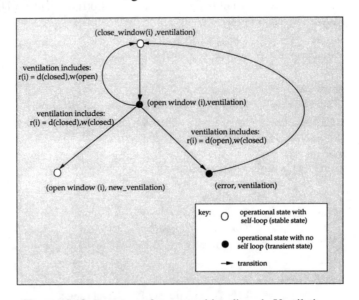

Figure 10.18: Fragment of state transition digraph, Ventilation

DE	DT ventilation(...r(i)...)	DE x DT ventilation(...r(i)...)	comments
close_window(i)	r(i)=(d(closed),w(closed))	close_window(i), r(i)=(d(closed),w(closed))	stable
open_window(i)	r(i) = (d(closed),w(open))	open_window(i), r(i)=(d(closed),w(closed))	transient
error	r(i) = (d(open),w(closed))	open_window(i), r(i) = (d(closed),w(open))	stable
		open_window(i), r(i) = (d(open),w(closed))	transient
		error(i),r(i)=(d(open),w(closed)	transient

Figure 10.19: Fragment of Part 1 of state table of Ventilation as shown in Figure 10.18

CURRENT STATE (DEc , DTc ventilation(...r(i)...))	NEXT STATE (DEn , DTn ventilation(...r(i)...))
close_window(i),r(i) = (d(closed),w(closed))	close_window(i),r(i) = (d(closed),w(closed))
open_window(i), (r(i)=d(closed),w(closed))	open_window(i), r(i)=(d(closed),w(open))
open_window(i), (r(i) = (d(closed),w(open))	close_window, r(i) = (d(closed),w(open))
open_window(i), (r(i) = (d(open),w(closed))	error, r(i) = (d(open),w(closed))
error, r(i) = (d(open),w(closed))	close_window(i), r(i) = (d(open),w(closed))

Figure 10.20: Fragment of Part 2 of forced transition table of Ventilation

CURRENT STATE	NEXT STATE
(_,(building,ventilation))	(open_window(i),(building,ventilation))

Figure 10.21: Fragment of Part 3 of free will transition table of Ventilation

10.4.3 The product of stage 3: Direct translations into Prolog

Apart from the first specification which is already expressed in Prolog, all the other four specifications were translated line by line into High Level Prolog as the common reference language. No account was taken of the operational nature of Prolog: it was used as a common *syntax*. Line-by-line equivalences are given in ([1]).

10.4.4 The product of stage 4: The expanded Full Prolog text

The text of each of the High Level Prolog translations can be expanded to Full Prolog by using the definitions for those members of the Mathematical Toolkit required by each individual specification. This is a routine procedure, which is facilitated by the measurement tool PDQ (see section 10.2.4).

10.4.5: The product of stage 5 : The measures

The code texts produced were checked for syntactic correctness with a syntax checker, and were then submitted individually as inputs to the two versions of PDQ described in Section 10.2.4. The first version produced the High Level Prolog measures by setting the Mathematical Toolkit measures to the identity measure for those definitions. The second version produced the measures for the equivalent Full Prolog versions of code text by substituting library values for the Mathematical Toolkit measures.

The measures were generated by PDQ in detail over all elements of the text hierarchy, building them up from the measures of predicates all the way to the measures of programs. The resulting program measures – the measures of the specifications – are presented in detail in Tables 10A.1 to 10A.11 of the APPENDIX to this chapter. In this section we show only an overview.

Figure 10.22 lists the top level six-tuple measures for both the High Level and Full Prolog for the four versions of the translated specification, together with the Full Prolog measure for the specification written directly in Prolog.

10.5 Observations

10.5.1 Conducting the experiment

We must recall that our purpose here is not to develop definitive measures for specifications, but to demonstrate multiple-language model-based measurement. Only a selection can be given here of the observations which emerged in carrying out each stage of the experiment.

Stage 1: Initially the specification of a ventilation system was prepared which arose from a real-life environmental control problem, but this was gradually slimmed down for presentation here, so as not to bloat

Original specification language	High Level Prolog	Full Prolog
Prolog	---	(29,17,18,13,13,1)
Z	(92,31,61,4,8,2)	(169,67,156,6,69,2)
VDM	(58,19,39,1,14,2)	(75,27,70,1,36,2)
Data Flow Diagrams (DFD)	(37,16,21,3,18,2).	(44,19,36,3,29,2).
State Transition Formalism (STF)	(22,11,15,10,22,2)	(24,13,20,12,27,2)

Figure 10.22: Overview of specification measures

the example. Key features (such as initial conditions and invariant) were retained.

Stage 2: To illustrate the potential scope of the multiple-language strategy, two formal and two semi-formal specification languages were selected for inclusion in the experiment, in addition to Prolog itself. The choice was guided by current interest in the selected media, and the ease with which instructions for translations into Prolog were readily available.

The 'ventilation' example was equally easy to specify in each language. The content of the specification depended on the problem itself, the rigour of the chosen medium (see e.g. the description of the Data Flow Diagram notation in Section 10.2.1), and the specifier's experience. Semi-formal specifications allow greater freedom to a specifier, but unfortunately this includes the freedom to make mistakes and omissions. Our expectation is that specifiers familiar with formal media would be more rigorous in their use of semi-formal syntax, for example, by inclusion of invariants. Wider experimental evidence would be needed to verify this.

Stage 3: The literature and our own experience shows that the translation of specifications into High Level Prolog causes very little difficulty. In addition to those chosen, translations have also been developed for some other specification notations. For example, a specification method based on Abstract Data Types has been developed at Philips in Brussels and at Bruges University, and a Prolog translation of a specification in this formalism is shown in ([13]). A specification language called Forest, reported in ([14]), has also been translated into Prolog for the purpose of animation. Although not considered in further detail here, these and other specification notations might readily be incorporated into our multiple-language strategy. Extension of the measurement scheme to parallel Prolog would allow measurement of concurrent specification media such as CCS ([15]).

Stage 4: This stage of the experiment revealed the problem of expressing the Mathematical Toolkit in Prolog. Various definitions are possible for each separate tool, and the literature offers options. For example, the procedure to define 'union' may be expressed declaratively, as in Code text 10A.7 of the Appendix, or procedurally as found in most Prolog textbooks. These will generate different measures. The full report on the experiment in ([1]) discusses various tool definitions at length, in particular 'setof' defined both as tester of set membership and as a generator of sets, and provides measures for these. The conclusion reached there is to give preference to a declarative definition which mirrors the accepted mathematical definition. Where several definitions exist for a given mathematical concept, as is the case for the example 'prime' furnished in ([1]), then the minimum measure should be accepted as the standard. Placing standard measures in the PDQ library allows the Prolog programmer freedom to use an operationally more efficient version of a mathematical tool, knowing that the measurement process will yield a unique specification measure, because any variation caused by implementation will be overridden by the standard measure incorporated in the PDQ library.

Stage 5: Initially the measures were calculated by inspection, but this proved error-prone and unreliable. To overcome this difficulty, the automated tool PDQ was developed. The tool itself was first produced for the measurement of Z specifications, but expansion to cover other representations was not difficult. Further expansion to other specification languages should pose no problem.

10.5.2 Observations on the measured results

Comparing High Level and Full Prolog specifications of a given laguage

This amounts to comparing entries in each *row* of Figure 10.22. The measured results are consistent with expectations: in each row, for each element of the 6-tuple, and for each of Z, VDM, DFD and STF:

High Level measures ≤ Full Prolog measures.

Measuring the levels of specification languages

We know that, in general, comparison between High Level and Full Prolog measures is indicative of the level of a language: the more the two differ, the higher is the level of the language. It would be far too hasty to devise a measure for our selected languages on the basis of a single example, but assume for the moment that we are willing to take the ventilation example as representative. Then how could we arrive at an aggregate measure on the basis of the 6-tuples of Figure 10.22 so that the four languages could be sorted in the order of their level?

We take two simplistic approaches to illustrate the way one may proceed. The reader is urged to bear in mind the scantiness of the data and the arbitrariness of the scale design: we are illustrating methods of measurement, rather than providing definitive measures for specification languages.

VERSION 1

- Add the figures within each 6-tuple .
- Subtract the sum for the High Level Prolog column from that of the Full Prolog column to give a measure of the extra 'Low Level Prolog' needed to implement the logic.
- Order the languages in accord with this number.

This gives:
$$STF <^L DFD <^L VDM <^L Z,$$
where $<^L$ signifies the relation 'lower-level than'.

END OF VERSION 1

VERSION 2

- Divide corresponding 6-tuple parameters for the row of Figure 10.22, taking the Full Prolog parameter as numerator and the High Level Prolog parameter as denominator.
- Add the results to gain a single number for each row.
- Order the languages accordingly.

This gives:
$$STF <^L DFD <^L VDM <^L Z,$$
which is the same result as for Version 1.

END OF VERSION 2

(The arithmetic is trivial and is not included.)

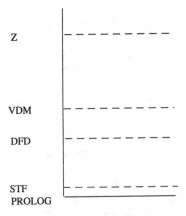

Figure 10.23: Uncritical comparison of the levels of specification languages

Of course, ordinal measures may be derived in other ways. We may also devise more sophisticated measurement methods, for example by attaching more significance to some 6-tuple parameters than to others, and we may also build a more discerning scale, instead of the ordinal scale above.

For the moment these results should be treated with caution. If an experiment were to be devised for measuring language levels on the basis of direct comparison of 6-tuples, such casual inspection of a table of measures would be insufficient. Only two of the aspects of experimental design are mentioned here:

- It would be necessary to define and validate an explicit indirect measure for LEVEL, in the form LEVEL = f(A, V, I, C, S, D), checking its consistency with all requirements of model-based measurement.

- One would have to assure that the information content of the measured text is consistent in all cases, and none of the specifications contain error, omission or redundancy. In the specifications in this experiment, rationalizing the Prolog code revealed a much greater element of redundancy for Z than for VDM, which inflates the Z measures. This is illustrated by examining, first, the Prolog definition of delta_ventilation (which will inflate the Z figures) and then the comment by the specifier of the VDM version of INIT.

10.5.3 Observations outside the experiment

Comparing referents – emergent properties

The experiment described here is directed at a single referent, and hence it can lead to no comparative conclusions about referents. However experiments reported elsewhere ([11, 16]) confirm an assumption that measures are related to the textual volume of a specification which in turn relates to rigour, and bear out the relative sizes of Z and VDM measures reported here. They also show how the model-based measurement scheme might be developed in future, in course of validating the measures of specifications.

As an example, we bring to the reader's attention some interesting results which have emerged from a much larger study of some independent translations of Z specifications, reported in ([16]). These results showed that over the hierarchy of the procedures of a single specification the ratios

$$\frac{V+C+I}{A+S}\,,\qquad\qquad \frac{I}{V+C+I}\,,\qquad\qquad \frac{I+S}{V+C+I}\,,$$

were remarkably consistent, while varying significantly from specification to specification. These indirect measures were first suggested in Chapter 9, and are shown here in Figure 10.24 for the Ventilation experiment. Observing the patterns, we speculate that these derived measures point to some emergent attributes which are as yet undefined, and will, in some way, reflect the nature of the specification. Measure validation experiments would be needed to confirm this.

Other factors affecting the measures

In Chapter 6 it was suggested that the proposed multiple language model-based measurement scheme could lead to the characterization of several factors of interest, but that the difficulty was to isolate their effects. The factors may affect the measures at any stage of the process from concept to final result, and several have already been mentioned in Section 10.5.1 in the context of the experiment. Further influences on the measures are now identified:

Rationalization of code. If, as in most cases, the translation of a specification into Prolog is carried out in order to animate the specification, the raw code will need to be converted into correct and efficient program code. This must take into account operational considerations. One of these is to define procedures that execute in the shortest time, especially those that execute most often, i.e. the Mathematical Toolkit procedures. Further, the original specification may contain redundancies, and these should be removed and the code reordered to give a shorter search tree in Prolog's quest for a solution. A single instance of such rationalization is shown here. It differentiates the code of tables 10A.1 and 10A.2, where the occurrence of list searching is reduced. Such a reordering may be essential to avoid the 'black hole' of a non-terminating search. These considerations, which will be familiar to the Prolog practitioner, are referred to as 'rationalization'.

High level Prolog	A+S	V+C+I	$\dfrac{V+C+I}{A+S}$	$\dfrac{I}{V+C+I}$	$\dfrac{C}{V+C+I}$	$\dfrac{I+S}{V+C+I}$
Z building (Table 10A.3)	100	96	.96	.64	.04	.72
VDM open window + invariants (comment on Table 10A.4)	72	59	.82	.66	.02	.90
DFD building + typing and invariants (comment on Table 10A.5)	55	40	.73	.53	.08	.98
State Transition building (Table 10A.6)	44	36	.82	.42	.28	1.03

Full Prolog	A+S	V+C+I	$\dfrac{V+C+I}{A+S}$	$\dfrac{I}{V+C+I}$	$\dfrac{C}{V+C+I}$	$\dfrac{I+S}{V+C+I}$
Prolog building (Table 10A.1)	42	48	1.14	.38	.27	.65
Revised Prolog building (Table 10A.2)	35	40	1.14	.38	.28	.68
Z building (Table 10A.7)	238	229	.96	.68	.03	.98
VDM open window1 +invariants (comment on Table 10A.8)	111	98	.88	.71	.01	1.08
DFD building (comment on Table 10A.9)	73	58	.79	.62	.05	1.12
State Transition building (Table 10A.10)	51	45	.88	.44	.27	1.04

Figure 10.24: Comparison of indirect measures

Monitoring the effects of rationalization is an additional use to which the measurement process may be put. Figure 10.25 shows how rationalization would be incorporated into the measurement process, leading to the expansion of Figure 10.22 by a whole new column of 6-tuple measures.

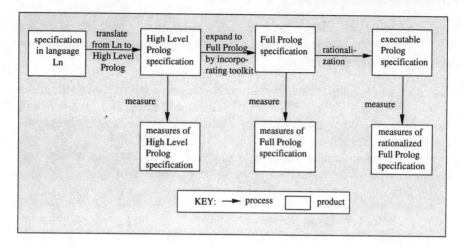

Figure 10.25: Inserting rationalization into the measurement process

Programming style also has an effect on the measures. Data may be passed through variables in the head of a clause or may be retrieved directly from the data base. There are effects caused by introduction of counters, guards, explicit identities, and replacement of structures by extra arguments. The magnitudes of the changes caused by these have received preliminary investigation. The results are reported in [1].

At this stage of development of a model-based measurement system for specification, only tentative suggestions can be made about isolating these factors and evaluating their individual effects. The validation or refutation of these conjectures is beyond the scope of this experiment, and would be the matter for further work in the development of the measurement scheme.

10.6 Conclusions

The aim of our case study was to demonstrate model-based measurement at work, and show the feasibility of applying it to measure specifications. The field of measurement and the design of the measures was chosen with care; however, the principles of model-based measurement are universal, and they could have been demonstrated on referents other than specifications, and measures other than the 6-tuple. In this method, the measures are not contrived ideas or arbitrary inventions: they arise naturally from the model. If the model is a valid representation of the referent then the measures tell a valid story about the referent. The story might even take unexpected turns, such as revealing attributes of which the modeller had been unaware. Such is the power of model-based measurement.

Although we have not sought to provide a definitive characterization for specifications, the 6-tuple turns out to be quite a promising practical tool for the engineer and the manager.

As an example, consider the managerial decision of electing a specification medium for a new project. If the team is newly assembled then members will be dedicated (and frequently dogmatic) followers of various different specification approaches. How is the manager to choose? He/she might consider three criteria: mathematical rigour, ease of implementation, and cost of development. To the last, the ordinal measure of 'level' serves as a guide.

Costing of software has sometimes been done on the basis of predicting the 'size' of code, comparing the new project with similar, completed projects. In the industry, lines-of-code (LOC) is taken as a standard measure. On a trivial level, there have been two well-known difficulties with its use: (i) how to define 'lines of code' as a standard unit of measurement, and (ii) how to discriminate between 'minor' lines (such as data declarations) and 'complex' lines of manipulation, which require creativity and a great deal of care. A more important objection to such an approach to cost prediction is that it cannot be applied to innovative projects, no comparative code being available, and even where programs of an apparently similar nature have been written before, they may conceal fundamental differences which undermine the whole basis of cost-prediction. The 6-tuple, being based on the specification, offers a key, requiring no bench-mark comparisons.

For a final example of the use of the 6-tuple, consider the well-known problem that the customer's requirements may change while the project is still under development – a reason for many a spectacular failure. To guard against the consequences of such changes, the project manager needs to adapt a strategy of 'least commitment': keeping the specification and design in generic form as far as possible along the development process (remember the discussion of generic specifications in Chapter 5). As a quality check of the 'specificity' of specifications and designs, the project manager may use the ratio C/V of the 6-tuple. Moreover, changes in the measures of the specification would provide a demonstrable basis for assessing additional charges.

Even though we talk of a 'final example', this account is merely a foretaste of the possible present and future uses of the simple 6-tuple. The potential of model-based measurement is limitless.

10.7 References

1   Myers M (1992):" Model-based measurement of formal and semi-formal specifications and experiments. Part 1: "Models of specifications, Part 2: Measures of specifications", Kaposi Associates Internal Report G23/1992 (available from the authors).

2   Hayes I, ed. (1987): "Specification case studies". Prentice Hall.

3   Jones, C B (1986): "Systematic Software Development using VDM." London. Prentice Hall International.

4   Yourdon E, Constantine L L (1979): "Structured design: Fundamentals of a Discipline of Computer Program and Systems design." Prentice Hall.

5   Knott R D, Krause P.J (1988): "An approach to animating Z using Prolog" Report No. A1.1, Alvey Project SE/065 Mathematics Department, Surrey University.

6   Knott R D (1992): "Implementation of Z specifications using program transformation systems: The Suzan Project". In "The Unified Computation Laboratory", pp 207-220 eds C.M.I Rattray & R.G Clark, the Institute of Mathematics and its Applications. Oxford University Press.

7   White A P (1983): "An evaluation of Prolog as a rapid prototyping language for formal specification", MSc dissertation. Manchester University.

8   Goble T (1989): "Structured systems analysis through Prolog". Prentice Hall.

9   Myers M (1990): "Quality assurance of specification and design of software". PhD dissertation. South Bank University.

10 Spivey J M (1989): "The Z Notation, A Reference Manual". Prentice Hall.

11 Benwood H J J (1988): "A software tool for the structural measurement of Prolog". MSc dissertation. Heriot-Watt University.

12 Birznieks R (1990): "Measurement of Specification using Prolog", MSc dissertation, South Bank University, London.

13 Habra N, van Lamsweerde A (1988): "Generating Prolog Prototypes from formal Specifications of Functional Requirements", 4th Conference:Exhibition Software Engineering. Paris (ISBN 2903677522).

14 Cunningham R, Finkelstein A, Goldsack S, Maibaum T, and Potts C (1986): "Formal Requirements Specification – The FOREST Project". Proc IEEE Third International Workshop on Specification and Design. pp 186-192. IEEE Comp Soc Press.

15 Milner R (1989): "Communication and Concurrency". Prentice Hall.

16 Myers M (1990): "Case studies and derived conclusion on measurement of Z". Contract ML316214 Deliverable No. 3, RT31, BT, Martlesham.

Appendix:

Prolog code and measures of the specifications

The Prolog specification

```prolog
valid_building(B):-
    forall(member(R,B),
    (R=[D,W],member(D,[open,closed]),member(W,[open,closed]))).

valid_ventilation(V):-
    forall((member(R,V), R=[D,W]),not (D=open,W=open)).

open_window(RoomNum,Building,NewBuilding):-
    valid_building(Building),
    valid_ventilation(Building),
    find_room(RoomNum,Building,Room),
    Room ==[closed,closed],                  % a condition of fulfilling
                                             %the request
    NewRoom==[closed,open],
    change_room(RoomNum,Building,NewRoom,NewBuilding).
open_window(RoomNum,Building,Building):-
    valid_building(Building),
    valid_ventilation(Building),
    find_room(RoomNum,Building,Room),
    Room =\= [closed,closed],
    Msg = 'cant do that',
    send(Msg).

send(Msg).

find_room(Num,List,Element):-
        find_room1(Num,0,List,Element).
                                        %introduce & initialize counter
find_room1(Num,Counter,[H|Tail],Room):-
    Counter1 is Counter + 1,
    Counter1 = Num,                      %room found
    Room = H.                            %give value to Room, end
find_room1(Num,Counter,[H|Tail],Room):-  %recursion
    Counter1 is Counter + 1,
    find_room1(Num,Counter1,Tail,Room).  %continue recursion
find_room1(Num,Counter,[],_).            %guard in case I too large

change_room(Num,List,NewElement,NewList):-
        change_room1(Num,0,List,NewElement,NewList).%introduce &
initialize counter
change_room1(Num,Counter,[_|Tail],NewElement,[NewElement|Tail]):-
        Counter1 is Counter+1,
        Counter1==Num.                   %test and terminate
change_room1(Num,Counter,[Head|Tail],NewElement,[Head|Tail1]):-
        Counter1 is Counter+1,
        change_room1(Num,Counter1,Tail,NewElement,Tail1).
```

Code text 10A.1: The Prolog Ventilation specification (contd. over)

```
member(H,[H|T]).
member(X,[H|T]):-member(X,T).
init_test(B):-
    forall(member(R,B),R==[closed,closed]).

init_gen(I,B):-
    init_gen1(I,0,[],B).
init_gen1(I, Counter,B,B):-
    Counter=I.                      %terminates recursion, instantiates B
init_gen1(I,Counter,PartB,B):-
    Counter1 is Counter + 1,
    init_gen1(I,Counter1,[[closed,closed]|PartB],B).%add next room
```

Code.text 10A.1 (contd.): The Prolog Ventilation specification

Clause	Ph	Pb	Pe	M1	M
valid_building/1/1	1,1,0,0,0,0	8,4,3,4,3,1	0,- 3,3,4,3,1	8,5,6,4,6,1	8,5,6,4,6,1
send/1/1	1,1,0,0,0,0	0,0,0,0,0,0	1,1,0,0,0,0	1,1,0,0,0,0	1,1,0,0,0,0
init_test/1/1	1,1,0,0,0,0	4,2,1,2,1,1	0,- 1,1,2,1,1	4,3,2,2,2,1	4,3,2,2,2,1
valid_ventilation/1/1	1,1,0,0,0,0	8,4,3,2,1,1	0,- 3,3,2,1,1	8,5,4,2,2,1	8,5,4,2,2,1
find_room1/4/1 find_room1/4/2 find_room1/4/3	4,5,0,0,1,1 4,5,0,0,1,1 4,3,0,1,0,0	4,4,0,0,0,0 4,4,0,0,0,0 0,0,0,0,0,0	0,0,0,0,1,1 0,0,0,0,1,1 4,3,0,1,0,0	4,4,0,0,1,1 4,4,0,0,1,1 4,3,0,1,0,0	4,4,0,1,1,1
change_room1/5/1 change_room1/5/2	5,5,2,0,2,1 5,6,1,0,2,1	2,2,0,0,0,0 5,5,0,0,0,0	0,0,2,0,2,1 0,0,1,0,2,1	2,2,2,0,2,1 5,5,1,0,2,1	5,5,2,0,2,1
init_gen1/4/1 init_gen1/4/2	4,3,1,0,0,0 4,4,0,0,0,0	2,2,0,0,0,0 4,4,0,2,2,2	0,0,1,0,0,0 0,0,0,2,2,2	2,2,1,0,0,0 4,4,0,2,2,2	4,4,0,2,2,2
find_room/3/1	3,3,0,0,0,0	4,3,0,1,0,0	0,0,0,1,0,0	4,4,0,2,1,1	4,4,0,2,1,1
member/2/1 member/2/2	2,2,1,0,1,1 2,3,0,0,1,1	0,0,0,0,0,0 2,2,0,0,0,0	2,2,1,0,1,1 0,0,0,0,1,1	2,2,1,0,1,1 2,2,0,0,1,1	2,2,1,0,1,1
init_gen/2/1	2,2,0,0,0,0	4,2,0,2,0,0	0,0,0,2,0,0	4,4,1,4,2,2	4,4,1,4,2,2
change_room/4/1	4,4,0,0,0,0	5,4,0,1,0,0	0,0,0,1,0,0	5,5,2,1,2,1	5,5,2,1,2,1
open_window/3/1 open_window/3/2	3,3,0,0,0,0 3,2,1,0,0,0	13,5,6,4,2, 1 8,4,3,1,0,0	0,- 6,6,4,2,1 0,- 3,4,1,0,0	29,17,18,13,13,1 23,14,14,9,9,1	**29,17,18,13,13,1**

Table 10A.1: Measurement of Prolog Ventilation specification

Comments on code text 10A.1 and results table 10A.1

The code is inefficient as list searching occurs twice, once in 'find_room/3', and once in 'change_room/3'.

The two independent procedures 'init_gen/1' and 'init_test/2' give the first means of comparing two procedures, one to test a state, and one to generate that same state.

Revised Prolog Specification

```
valid_building(B):-
    forall(member(R,B),
    (R=[D,W],member(D,[open,closed]),member(W,[open,closed]))).
valid_ventilation(V):-
        forall((member(R,V),R=[D,W]),not(D=open,W=open)).
open_window(RoomNum,Building,NewBuilding):-
    valid_building(Building),
    valid_ventilation(Building),
    find_change(RoomNum,Building,NewBuilding).

find_change(RoomNum,Building,NewBuilding):-
    find_change1(RoomNum,0,Building,NewBuilding).
                                    %add counter
find_change1(RoomNum,Counter,[Room|Rest],[NewRoom|Rest]):-
    Counter1 is Counter +1,
    Counter1 = RoomNum,
    Room ==[closed,closed],        %test
    NewRoom=[closed,open].         %assign and terminate recursion
find_change1(RoomNum,Counter,[Room|Rest],[Room|Rest]):-
    Counter1 is Counter +1,
    Counter1 = RoomNum,
    Room \==[closed,closed],
    Msg = 'illegal operation',
    send(Msg).                     %terminate recursion with failure
find_change1(RoomNum,Counter,[Room|Rest],[Room|Rest1]):-
    Counter1 is Counter +1,
    find_change1(RoomNum,Counter1,Rest,Rest1).
                                    %continue recursive search
find_change1(_,_,[],_):-
    Msg = 'No such room',
    send(Msg).                     %needed if no previous check on
                                    %magnitude of RoomNum

send(Msg).

member(H,[H|T])).
member(X,[H|T]):-member(X,T).
init_test(B):-
    forall(member(R,B),R==[closed,closed]).
```

Code text 10A.2 Revised Prolog Building specification, partially rationalized

```
init_gen(I,B):-
    init_gen1(I,0,[],B).
init_gen1(I,Counter,B,B):-
    Counter=I.                      %terminates recursion,
                                    %instantiates B
init_gen1(I,Counter,PartB,B):-
    Counter1 is Counter + 1,
    init_gen1(I,Counter1,[[closed,closed]|PartB],B).
                                    %add next room
```

Code text 10A.2 (cont.): Revised Prolog Building specification, partially rationalized

Clause	Ph	Pb	Pe	M1	
send/1/1	1,1,0,0,0,0	0,0,0,0,0,0	1,1,0,0,0,0	1,1,0,0,0,0	1,1,0,0,0,0
member/2/1	2,2,1,0,1,1	0,0,0,0,0,0	2,2,1,0,1,1	2,2,1,0,1,1	
member/2/2	2,3,0,0,1,1	2,2,0,0,0,0	0,0,0,0,1,1	2,2,0,0,1,1	2,2,1,0,1,1
init_gen/2/1	2,2,0,0,0,0	4,2,0,2,0,0	0,0,0,2,0,0	4,4,1,4,2,2	4,4,1,4,2,2
init_gen1/4/1	4,3,1,0,0,0	2,2,0,0,0,0	0,0,1,0,0,0	2,2,1,0,0,0	
init_gen1/4/2	4,4,0,0,0,0	4,4,0,2,2,2	0,0,0,2,2,2	4,4,0,2,2,2	4,4,1,2,2,2
valid_building/1/1	1,1,0,0,0,0	8,4,3,4,3,1	0,- 3,3,4,3,1	8,5,6,4,6,1	8,5,6,4,6,1
find_change1/4/1	4,5,1,0,2,1	6,4,0,4,2,1	0,0,1,4,4,1	6,6,1,4,4,1	
find_change1/4/2	4,4,2,0,2,1	7,4,1,3,1,1	0,- 1,3,3,3,1	7,6,3,3,3,1	
find_change1/4/3	4,5,1,0,2,1	4,4,0,0,0,0	0,0,1,0,2,1	4,4,1,0,2,1	
find_change1/4/4	4,3,0,1,0,0	3,1,1,1,0,0	0,- 1,1,2,0,0	3,2,1,2,0,0	7,6,3,4,4,1
valid_ventilation/1/1	1,1,0,0,0,0	8,4,3,2,1,1	0,- 3,3,2,1,1	8,5,4,2,2,1	8,5,4,2,2,1
find_change/3/1	3,3,0,0,0,0	4,3,0,1,0,0	0,0,0,1,0,0	7,6,3,5,4,1	7,6,3,5,4,1
open_window/3/1	3,3,0,0,0,0	5,3,2,0,0,0	0,- 2,2,0,0,0	23,14,15,11,12,1	**23,14,15,11,12,1**
init_test/1/1	1,1,0,0,0,0	4,2,1,2,1,1	0,- 1,1,2,1,1	4,3,2,2,2,1	4,3,2,2,2,1

Table 10A.2: Measure of revised Prolog Building

Comment on code text 10A.2 and table 10A.2

The inefficiency noted above has been removed. Comparison between the measures of the first Prolog Building text and the revised version shows a significant drop. In this case the more efficient code has a lower measure.

The Z specification

```
mybuilding(Doors,Windows,Orifice,Building):-
    powerset(Orifice,POrifice),
    member(Doors,POrifice), member(Windows, POrifice),
    is_a_relation(Doors,Windows,Rooms),
    isZseq(Building), rng(Building, Rooms),
    intersect(Doors,Windows,[]),

forall((ithofZseq(I,Building,[Di,Wi]),ithofZseq(J,Building,[Dj,Wj])),
                        ((not Di=Dj ; I=J),(not Wi=Wj ; I=J))).

ventilation(Doors,Windows,Orifice,Building,Open_set,Closed_set):-
    mybuilding(Doors,Windows,Orifice,Building),
    powerset(Orifice,POrifice),
    member(Open_set,POrifice),member(Closed_set,POrifice),
    intersect(Open_set,Closed_set,[]),

forall(member(O,Orifice),(member(O,Open_set);member(O,Closed_set))).

init(Doors,Windows,Orifice,Building,Open_set,Closed_set):-
    ventilation(Doors,Windows,Orifice,Building,Open_set ,Closed_set),
    Open_set = [],
    Closed_set = Orifice.

delta_ventilation(Doors,Windows,Orifice,Building,Open_set,Closed_set,

Open_setp, Closed_setp):-
    ventilation(Doors,Windows,Orifice,Building,Open_set,Closed_set),
    ventilation(Doors,Windows,Orifice,Building,Open_setp,Closed_setp).

windowOpenOK(I,Doors,Windows,Orifice, Building, Open_set,
                            Closed_set,Open_setp, Closed_setp):-
    delta_ventilation(Doors,Windows,Orifice,Building,Open_set,
                            Closed_set,Open_setp,Closed_setp),
    dom(Building,Dom),member(I,Dom),
    member([I,[DoorI,WindowI]],Building),
    member(WindowI,Closed_set),member(DoorI,Closed_set),
    set_difference(Closed_set,[WindowI],Closed_setp),
    union(Open_set,[Window],Open_setp).

xi_ventilation(Doors,Windows,Orifice,Building,Open_set,Closed_set,
                            Open_setp, Closed_setp):-
    Open_set=Open_setp,Closed_set=Closed_setp.

openErr(I,Doors,Windows,Orifice,Building,Open_set,Closed_set,
                            Open_setp,Closed_setp):-
    xi_ventilation(Doors,Windows,Orifice,Building,Open_set,Closed_set,
                            Open_setp, Closed_setp),
    dom(Building,Dom),member(I,Dom),
    member([I,[DoorI,WindowI]],Building),
    (member(WindowI,Open_set);member(DoorI,Open_set)),
    send('already vented').
```

Code text 10A.3: The Z building (contd. over)

```
open_window(I,Doors,Windows,Orifice,Building,Open_set,Closed_set,
Open_setp,Closed_setp):-
    windowOpenOK(I,Doors,Windows,Orifice,Building, Open_set,
                             Closed_set,Open_setp,Closed_setp).

open_window(I,Doors,Windows,Orifice,Building,Open_set,Closed_set,
                              Open_setp,Closed_setp):-
    openErr(I,Doors,Windows,Orifice,Building, Open_set, Closed_set,
                              Open_setp, Closed_setp).

send(Msg).
```

Code text 10A.3 (contd.): The Z building

Clause	Ph	Pb	Pe	M1	M
xi_ventilation/8/1	8,8,0,0,0,0	4,4,0,0,0,0	0,0,0,0,0,0	4,4,0,0,0,0	4,4,0,0,0,0
send/1/1	1,1,0,0,0,0	0,0,0,0,0,0	1,1,0,0,0,0	1,1,0,0,0,0	1,1,0,0,0,0
mybuilding/4/1	4,4,0,0,0,0	25,12,14,1,2,1	0,14,14,1,2,1	25,11,14,1,2,1	
mybuilding/4/2	4,4,0,0,0,0	25,12,14,1,2,1	0,14,14,1,2,1	25,11,14,1,2,1	
mybuilding/4/3	4,4,0,0,0,0	25,12,14,1,2,1	0,14,14,1,2,1	25,11,14,1,2,1	
mybuilding/4/4	4,4,0,0,0,0	25,12,14,1,2,1	0,14,14,1,2,1	25,11,14,1,2,1	25,11,14,1,2,1
openErr/9/1	9,9,0,0,0,0	17,12,6,1,2,2	0,- 6,6,1,2,2	13,7,6,1,2,2	
openErr/9/2	9,9,0,0,0,0	17,12,6,1,2,2	0,- 6,6,1,2,2	13,7,6,1,2,2	13,7,6,1,2,2
ventilation/6/1	6,6,0,0,0,0	17,8,8,1,0,0	0,-8,8,1,0,0	38,16,22,2,2,1	
ventilation/6/2	6,6,0,0,0,0	17,8,8,1,0,0	0,-8,8,1,0,0	38,16,22,2,2,1	38,16,22,2,2,1
delta_ventilation/8/1	8,8,0,0,0,0	12,8,4,0,0,0	0,-4,4,0,0,0	76,28,48,4,4,1	76,28,48,4,4,1
init/6/1 6,6,0,0,0,0	6,6,0,0,0,0	10,6,3,1,0,0	0,-3,3,1,0,0	42,17,25,3,2,1	42,17,25,3,2,1
windowOpenOK/9/1	9,9,0,0,0,0	24,13,13,0,4,2	0,-13,13,0,4,2	92,31,61,4,8,2	92,31,61,4,8,2
open_window/9/1	9,9,0,0,0,0	9,9,0,0,0,0	0,0,0,0,0,0	92,31,61,4,8,2	
open_window/9/2	9,9,0,0,0,0	9,9,0,0,0,0	0,0,0,0,0,0	13,7,6,1,2,2	**92,31,62,4,8,2**

Table 10A.3: Measures for HL Prolog of program Z building.text

Comments on code 10A.3 and table 10A.3

The measures of table 10A.3 are by far the highest of any of the specifications.

The introduction of the Δ and Ξ specifications in the form used immediately doubles the measures. For instance, 'delta_ventilation/8' contains two body predicates, each of which calls the type checker, 'mybuilding/4'. This is not the only situation in which redundancy appears. It may do so wherever there is a call to what was, in the original, a hierarchy of subordinate schemas. A reference to just this aspect has been investigated ([5]), and the authors of that paper have developed an automated program transformation scheme to eliminate redundancy caused in this way.

As in code text 10A.1, the acceptance condition, 'init/6' is separate from the operation hierarchy. It is a test of the required condition, but is wasteful in that it first type checks building, then checks the ventilation condition and then imposes yet a further constraint. This is apparent in the Prolog code, and the relevant measures draw immediate attention to it.

The VDM specification

```
type_building:-
    building(Building),                          %fetch value
    Building= [Rooms,Ventilation],
    Ventilation = [Opn,Clsd],
    orifices(Orifice_set),                       %fetch value
    subset(Opn,Orifice_set),subset(Clsd,Orifice_set),
    forall(member(Room,Rooms),
        (Room = [W,D],
            member(W,Orifice_set),member(D,Orifice_set))
        ).

orifices(Orifice_set).
building(Building).

inv_Ventilation:-
    building(Building),                          %fetch value
    Building =[_, [Opn,Clsd]],
    intersect(Opn,Clsd,[]).

inv_Building:-
    building(Building),
    Building = [Rooms,[Opn,Clsd]],
    forall(member(R,Rooms),not subset(R,Opn)),
    forall((member(R,Rooms),member(S,Rooms),R=[DR,WR],S=[DS,WS]), %gen-
    erate
        (not subset([DR,WR],Opn),                %test
        DR \== WS,                               % no sharing
        (DS \== DR ; S ==R),                     % door unique
        (WS \== WR ; S ==R))).                   % window unique
init:-
    building(B),
    Building=[Rms,Vent],
    Vent=[Opn,Clsd],
    forall(member(R,Rms), subset(R,Clsd)).
open_window(Room) :-
    building([Rooms, [Opn,Clsd]]),               % external read
```

Code text 10A.4: VDM building (contd. over)

```
        member(Room,Rooms),                        % precondition
        subset(Room,Clsd),                         % precondition
        union(Opn,[W],Opnp),                       % postcondition
        setdiff(Clsd,[W],Clsdp),           % implied postcondition
        retract(building(_)),                      % external write
        assert(building([Rooms,[Opnp,Clsdp]]))).
err_open_w(Room, Msg):-
        building([Rooms,[Opn,Clsd]]),              % external read
        member(Room,Rooms),                        % precondition
        not subset(Room,Clsd),                     % precondition
        send(Msg).                                 % postcondition
open_window1(Room):-
        open_window(Room).
open_window1(Room):-
        err_open_w(Room, Msg).
```

Code text 10A.4 (contd.): VDM building

Clause	Ph	Pb	Pe	M1	M
orifices/1/1	1,1,0,0,0,0	0,0,0,0,0,0	1,1,0,0,0,0	1,1,0,0,0,0	1,1,0,0,0,0
send/1/1	1,1,0,0,0,0	0,0,0,0,0,0	1,1,0,0,0,0	1,1,0,0,0,0	1,1,0,0,0,0
building/1/1	1,1,0,0,0,0	0,0,0,0,0,0	1,1,0,0,0,0	1,1,0,0,0,0	1,1,0,0,0,0
init/0/1	0,0,0,0,0,0	9,7,4,0,2,1	0,- 4,4,0,2,1	9,5,4,0,2,1	9,5,4,0,2,1
err_open_w/2/1	2,2,0,0,0,0	6,5,3,0,2,2	0,- 3,3,0,2,2	6,3,3,0,2,2	6,3,3,0,2,2
type_building/0/1	0,0,0,0,0,0	18,9,12,0,3,1	0,- 12,12,0,3,1	18,6,12,0,3,1	18,6,12,0,3,1
open_window/1/1	1,1,0,0,0,0	11,7,6,0,4,2	0,- 6,6,0,4,2	11,5,6,0,4,2	11,5,6,0,4,2
inv_ventilation/0/1	0,0,0,0,0,0	6,4,3,1,2,2	0,- 3,3,1,2,2	6,3,3,1,2,2	6,3,3,1,2,2
open_window1/1/1	1,1,0,0,0,0	1,1,0,0,0,0	0,0,0,0,0,0	11,5,6,0,4,2	
open_window1/1/2	1,1,0,0,0,0	2,2,0,0,0,0	0,-0,0,0,0,0	6,3,3,0,2,2	11,5,6,0,4,2
inv_building/0/1	0,0,0,0,0,0	23,10,18,0,5,2	0,- 18,18,0,5,2	23,5,18,0,5,2	
inv_building/0/2	0,0,0,0,0,0	23,10,18,0,5,2	0,- 18,18,0,5,2	23,5,18,0,5,2	
inv_building/0/3	0,0,0,0,0,0	23,10,18,0,5,2	0,- 18,18,0,5,2	23,5,18,0,5,2	
inv_building/0/4	0,0,0,0,0,0	23,10,18,0,5,2	0,- 18,18,0,5,2	23,5,18,0,5,2	23,5,18,0,5,2

Table 10A.4: Measures for HL Prolog of program VDM building

Comments on table 10A.4

The hierarchy generated by the 'open_window1/1' procedure does not include the type check or constraint check of procedures 'type_building/0', 'inv_building/0' or 'inv_ventilation/0', and thus is not directly comparable with the previous specifications. We can overcome this by introducing a top-level procedure not directly present in the original specification. This might be:

```
top_level(Room):-
  type_building,
  inv_building,
  inv_ventilation,
  open_window1(Room).
```

This procedure generates a measure of **(58,19,39,1,14,2)**.

Data Flow Diagram specification

```
type_building(B):-
    forall(member(R,B), R=[d(I),w(I)]),
    forall((member([d(I),w(I)],B),
            member([d(J),w(J)],B)),
                not I=J).
type_ventilation(O,C):-
    intersect(O,C,[]),
    forall((I, member([d(I),w(I)],O), not subset([door(I),window(I)],O))).

type_ventilation(O,C):-
    intersect(O,C,[]),
    forall(I, not(subset([door(I),window(I)],O))).

checked_state(Room,Result):-
    building(B),                    %get store
    member([d(Room),w(Room)], B),   %does the room exist?
    ventilation(Openset,Closedset), %get current state
    member(d(Room), Closedset),member(w(Room),Closedset),
                                    % should the window be opened?
    Result =ok.

checked_state(Room,Result):-
    building(B),                    %get store
    ventilation(Openset,Closedset), %get current state
    not (member([d(Room),w(Room)], B),%does the room exist?
            member(d(Room),Closedset),member(w(Room),Closedset)),
                                    % should the window be opened?
    Result =error.

updated_ventilation(Room):-
    ventilation(Openset,Closedset),
    union(Openset, [w(Room)], NewOpenset),
    set_difference(Closedset, [w(Room)], NewClosedset),
    retract(ventilation(_,_)),
    assert(ventilation(NewOpenset,NewClosedset)).   %update store.
openWresponse(Room):-
    checked_state(Room,Response),
    Response ==ok,
    updated_ventilation(Room).
```

Code text 10A.5: DFD building (contd. over)

```
openWresponse(Room):-
    checked_state(Room,Response),
    Response = error,
    send(Response).

building(B).
ventilation(O,C).
send(Msg).
```

Code text 10A.5 (contd.): DFD building

Clause	Ph	Pb	Pe	M1	M
type_building/1/1	1,1,0,0,0,0	10,4,8,0,9,2	0,- 8,8,0,9,2	10,2,8,0,9,2	10,2,8,0,9,2
building/1/1	1,1,0,0,0,0	0,0,0,0,0,0	1,1,0,0,0,0	1,1,0,0,0,0	1,1,0,0,0,0
send/1/1	1,1,0,0,0,0	0,0,0,0,0,0	1,1,0,0,0,0	1,1,0,0,0,0	1,1,0,0,0,0
type_ventilation/2/1	2,2,0,0,0,0	3,2,0,1,0,0	0,0,0,1,0,0	3,3,0,1,0,0	3,3,0,1,0,0
ventilation/2/1	2,2,0,0,0,0	0,0,0,0,0,0	2,2,0,0,0,0	2,2,0,0,0,0	2,2,0,0,0,0
updated_ventilation/1/1	1,1,0,0,0,0	8,5,3,0,4,2	0,- 3,3,0,4	8,5,3,0,4,2	8,5,3,0,4,2
checked_state/2/1	2,2,0,0,0,0	11,5,5,1,5,2	0,- 5,5,1,5,2	11,6,5,1,5,2	
checked_state/2/2	2,2,0,0,0,0	11,5,5,1,5,2	0,- 5,5,1,5,2	11,6,5,1,5,2	11,6,5,1,5,2
openWresponse/1/1	1,1,0,0,0,0	5,2,2,1,0,0	0,- 2,2,1,0,0	21,11,10,2,9,	
openWresponse/1/2	1,1,0,0,0,0	4,2,1,1,0,0	0,- 1,1,1,0,0	214,7,7,2,5,2	21,11,10,2,9,2

Table 10A.5: Measures for HL Prolog of program DFD building.

Comments on Table 10A.5

This specification, as the previous one, does not include type checking on building or venitilation in the top level procedure, openWresponse/1. The measure is thus comparable to that of table 10A.4. Again we can construct a composite, purely in order to obtain a measure for comparison with the top level of other specifications:

```
toplevel(Room):-
    building(B),
    type_building(B),
    ventilation(O,C),
    type_ventilation(O,C),
    openWresponse(Room).
```

The measure of this procedure, hierarchically composed over the rest of the text is **(37,16,21,3,18,2)**.

State transition specification

```
type_building([]).
type_building([Room|RestOfRooms]):-   Room = r(D,W),
                       member(D,[open,closed]),
                       member(W,[open,closed]),
                       type_building(RestOfRooms).


type_ventilation([]).
type_ventilation([Room|RestOfRooms]):-Room = r(D,W),
                       member(D,[open,closed]),
                       member(W,[open,closed]),
                       allowed(D,W),
                       type_ventilation(RestOfRooms).
allowed(D,W):- not (D=open,W=open).

% next is unfolded version of previous.
type_ventilation([Room|RestOfRooms]):-Room = r(D,W),
                       member(D,[open,closed]),
                       member(W,[open,closed]),
                       not (D=open,W=open),
                       type_ventilation(RestOfRooms).

forced([close_window(I),V],[close_window(I),V1]):-      %stable state
                       nthofList(I,V,r(closed,closed))
                       V1=V.
forced([open_window(I),V],[open_window(I),V1]):-
                       nthofList(I,V,r(closed,closed)),  %test
                       override(I,V,r(closed,open),V1).  %create
forced([open_window(I),V],[open_window(I),V1]):-
                       nthofList(I,V,r(closed,open)),    %test
                       V1 = V.                           %equate
forced([open_window(I),V],[error(I),V1]):-
                       nthofList(I,V,r(open,closed)),    %test
                       signal(error).
forced([error(I),V],[close_window(I),V1]):-
                       nthofList(I,V,r(open,closed)),
                       V1=V.
free_will([close_window(I),V],[open_window(I),V]).

signal(_).
open_window(I,V):-
     free_will([close_window(I), V],[open_window(I), V]),
                                      %is this a valid command?
     type_ventilation(V),                      %type check
     chain([open_window(I),V],[open_window(I),V1]). %launch response

chain(Current,Current):- forced(Current,Current).   %stable state found
chain(Current,Next):-
     forced(Current,New),
     chain(New,Next).
```

Code text 10A.6 : STF building

Clause	Ph	Pb	Pe	Ml	M
allowed/2/1	2,2,0,0,0,0	4,2,0,2,0,0	0,0,0,2,0,0	4,4,0,2,0,0	4,4,0,2,0,0
signal/1/1	1,1,0,0,0,0	0,0,0,0,0,0	1,1,0,0,0,0	1,1,0,0,0,0	1,1,0,0,0,0
type_building/1/1	1,0,0,1,0,0	0,0,0,0,0,0	1,0,0,1,0,0	1,0,0,1,0,0	
type_building/1/2	1,2,0,0,1,1	7,4,2,4,3,1	0,- 2,2,4,4,1	7,5,2,4,4,1	7,5,2,4,4,1
type_ventilation/1/1	1,0,0,1,0,0	0,0,0,0,0,0	1,0,0,1,0,0	1,0,0,1,0,0	
type_ventilation/1/2	1,2,0,0,1,1	9,4,4,4,3,1	0,- 4,4,4,4,1	11,7,4,6,4,1*	
type_ventilation/1/3	1,2,0,0,1,1	11,4,4,6,3,1	0,- 4,4,6,4,1	11,7,4,6,4,1*	11,7,4,6,4,1
free_will/2/1	2,2,2,0,4,2	0,0,0,0,0,0	2,2,2,0,4,2	2,2,2,0,4,2	2,2,2,0,4,2
forced/2/1	2,2,2,0,4,2	3,2,0,2,1,1	0,0,2,2,5,2	5,4,2,2,5,2	
forced/2/2	2,3,1,0,4,2	7,3,2,4,2,1	0,- 2,3,4,6,2	7,5,3,4,6,2	
forced/2/3	2,3,1,0,4,2	5,3,1,2,1,1	0,- 1,2,2,5,2	5,4,2,2,5,2	
forced/2/4	2,2,2,0,4,2	4,2,0,3,1,1	0,0,2,3,5,2	4,4,1,3,5,2	
forced/2/5	2,3,1,0,4,2	5,3,1,2,1,1	0,- 1,2,2,5,2	5,4,2,2,5,2	7,5,3,4,6,2
chain/2/1	2,1,1,0,0,0	2,1,1,0,0,0	0,- 1,2,0,0,0	7,4,5,4,6,2	
chain/2/	2,2,0,0,0,0	4,3,1,0,0,0	0,- 1,1,0,0,0	9,6,4,4,6,2	9,6,5,4,6,2
open_window/2/1	2,2,0,0,0,0	5,3,4,0,8,2	0,- 4,4,0,8,2	22,12,15,10,22,2	22,11,15,10,22,2

* same results since clause 3 is unfolded version of clause 2

Table 10A.6: Measures for HL Prolog of program State Transition building

Comments on Table 10A.6

The open_window top level procedure includes type checking on the ventilation which itself contains an implicit type check on the building. This is a more frugal version than others.

Two clauses of type_ventilation were included to show that measures are invariant over unfolding.

Clause	Ph	Pb	Pe	M1	M
xi_ventilation/8/1	8,8,0,0,0,0	4,4,0,0,0,0	0,0,0,0,0,0	4,4,0,0,0,0	4,4,0,0,0,0
send/1/1	1,1,0,0,0,0	0,0,0,0,0,0	1,1,0,0,0,0	1,1,0,0,0,0	1,1,0,0,0,0
mybuilding/4/1	4,4,0,0,0,0	25,12,14,1,2,1	0,-14,14,1,2,1	53,24,44,2,20,1	
mybuilding/4/2	4,4,0,0,0,0	25,12,14,1,2,1	0,-14,14,1,2,1	53,24,44,2,20,1	
mybuilding/4/3	4,4,0,0,0,0	25,12,14,1,2,1	0,-14,14,1,2,1	53,24,44,2,20,1	
mybuilding/4/4	4,4,0,0,0,0	25,12,14,1,2,1	0,-14,14,1,2,1	53,24,44,2,20,1	53,24,44,2,20,1
openErr/9/1	9,9,0,0,0,0	17,12,6,1,2,2	0,- 6,6,1,2,2	16,8,12,1,7,2	
openErr/9/2	9,9,0,0,0,0	17,12,6,1,2,2	0,- 6,6,1,2,2	16,8,12,1,7,2	16,8,12,1,7,2
ventilation/6/1	6,6,0,0,0,0	17,8,8,1,0,0	0,- 8,8,1,0,0	73,33,63,3,28,1	
ventilation/6/2	6,6,0,0,0,0	17,8,8,1,0,0	0,- 8,8,1,0,0	73,33,63,3,28,1	73,33,63,3,28,1
delta_ventilation/8/1	8,8,0,0,0,0	12,8,4,0,0,0	0,- 4,4,0,0,0	146,62,130,6,56,1	146,62,130,6,56,1
init/6/1	6,6,0,0,0,0	10,6,3,1,0,0	0,- 3,3,1,0,0	77,34,66,4,28,1	77,34,66,4,28,1
windowOpenOK/9/1	9,9,0,0,0,0	24,13,13,0,4,2	0,-13,13,0,4,2	169,67,156,6,69,2	169,67,156,6,69,2
open_window/9/1	9,9,0,0,0,0	9,9,0,0,0,0	0,0,0,0,0,0	169,67,156,6,69,2	
open_window/9/2	9,9,0,0,0,0	9,9,0,0,0,0	0,0,0,0,0,0	16,8,12,1,7,2	**169,67,156,6,69,2**

Table 10A.7: Measures for Full program Z building

Comment on table 10A.7

Again, very high values are generated by the Z specification language.

Clause	Ph	Pb	Pe	M1	M
orifices/1/1	1,1,0,0,0,0	0,0,0,0,0,0	1,1,0,0,0,0	1,1,0,0,0,0	1,1,0,0,0,0
send/1/1	1,1,0,0,0,0	0,0,0,0,0,0	1,1,0,0,0,0	1,1,0,0,0,0	1,1,0,0,0,0
building/1/1	1,1,0,0,0,0	0,0,0,0,0,0	1,1,0,0,0,0	1,1,0,0,0,0	1,1,0,0,0,0
init/0/1	0,0,0,0,0,0	9,7,4,0,2,1	0,- 4,4,0,2,1	11,6,8,0,5,1	11,6,8,0,5,1
err_open_w/2/1	2,2,0,0,0,0	6,5,3,0,2,2	0,-3,3,0,2,2	8,4,7,0,5,2	8,4,7,0,5,2
type_building/0/1	0,0,0,0,0,0	8,9,12,0,3,1	0,- 12,12,0,3,1	22,8,21,0,10,1	22,8,21,0,10,1
open_window/1/1	1,1,0,0,0,0	11,7,6,0,4,2	0,- 6,6,0,4,2	17,7,16,0,10,2	17,7,16,0,10,2
inv_Ventilation/0/1	0,0,0,0,0,0	6,4,3,1,2,2	0,- 3,3,1,2,2	9,5,6,1,4,2	9,5,6,1,4,2
open_window1/1/1	1,1,0,0,0,0	1,1,0,0,0,0	0,0,0,0,0,0	17,7,16,0,10,2	
open_window1/1/2	1,1,0,0,0,0	2,2,0,0,0,0	0,0,0,0,0,0	8,4,7,0,5,2	17,7,16,0,10,2
inv_Building/0/1	0,0,0,0,0,0	23,10,18,0,5,2	0,- 18,18,0,5,2	27,7,27,0,12,2	
inv_Building/0/2	0,0,0,0,0,0	23,10,18,0,5,2	0,- 18,18,0,5,2	27,7,27,0,12,2	
inv_Building/0/3	0,0,0,0,0,0	23,10,18,0,5,2	0,- 18,18,0,5,2	27,7,27,0,12,2	
inv_Building/0/4	0,0,0,0,0,0	23,10,18,0,5,2	0,- 18,18,0,5,2	27,7,27,0,12,2	27,7,27,0,12,2

Table 10A.8: Measures for Full Prolog of program VDM building (see comment)

Comment on table 10A.8

The corresponding measure for a 'top_level/0' procedure mentioned in the comments on table 10A.4 is **(75,27,70,1,36,2).**

Clause	Ph	Pb	Pe	M1	M
type_building/1/1	1,1,0,0,0,0	10,4,8,0,9,2	0,- 8,8,0,9,2	10,2,11,0,12,2	10,2,11,0,12,2
building/1/1	1,1,0,0,0,0	0,0,0,0,0,0	1,1,0,0,0,0	1,1,0,0,0,0	1,1,0,0,0,0
ventilation/2/1	2,2,0,0,0,0	0,0,0,0,0,0	2,2,0,0,0,0	2,2,0,0,0,0	2,2,0,0,0,0
send/1/1	1,1,0,0,0,0	0,0,0,0,0,0	1,1,0,0,0,0	1,1,0,0,0,0	1,1,0,0,0,0
type_ventilation/2/1	2,2,0,0,0,0	3,2,0,1,0,0	0,0,0,1,0,0	6,5,3,1,2,1	6,5,3,1,2,1
updated_ventilation/1/1	1,1,0,0,0,0	8,5,3,0,4,2	0,- 3,3,0,4,2	12,6,9,0,7,2	12,6,9,0,7,2
checked_state/2/1 checked_state/2/2	2,2,0,0,0,0 2,2,0,0,0,0	11,5,5,1,5,2 11,5,5,1,5,2	0,- 5,5,1,5,2 0,- 5,5,1,5,2	11,6,8,1,8,2 11,6,8,1,8,2	 11,6,8,1,8,2
openWresponse/1/1 openWresponse/1/2	1,1,0,0,0,0 1,1,0,0,0,0	5,2,2,1,0,0 4,2,1,1,0,0	0,- 2,2,1,0,0 0,- 1,1,1,0,0	25,12,19,2,15,2 14,7,10,2,8,2	 25,12,19,2,15,2

Table 10A.9 : Measures for Full Prolog of program DFD building (see comment)

Comment on table 10A.9

The measure of an added top level procedure for this specification, as mentioned in the comments on table 10A.5, is **(44,19,36,3,29,2)**.

Clause	Ph	Pb	Pe	M1	M
allowed/2/1	2,2,0,0,0,0	4,2,0,2,0,0	0,0,0,2,0,0	4,4,0,2,0,0	4,4,0,2,0,0
signal/1/1	1,1,0,0,0,0	0,0,0,0,0,0	1,1,0,0,0,0	1,1,0,0,0,0	1,1,0,0,0,0
type_building/1/1 type_building/1/2	1,0,0,1,0,0 1,2,0,0,1,1	0,0,0,0,0,0 7,4,2,4,3,1	1,0,0,1,0,0 0,- 2,2,4,4,1	1,0,0,1,0,0 7,5,4,4,6,1	 7,5,4,4,6,1
type_ventilation/1/1 type_ventilation/1/2 type_ventilation/1/3	1,0,0,1,0,0 1,2,0,0,1,1 1,2,0,0,1,1	0,0,0,0,0,0 9,4,4,4,3,1 11,4,4,6,3,1	1,0,0,1,0,0 0,- 4,4,4,4,1 0,- 4,4,6,4,1	1,0,0,1,0,0 11,7,6,6,6,1 11,7,6,6,6,1	 11,7,6,6,6,1
free_will/2/1	2,2,2,0,4,2	0,0,0,0,0,0	2,2,2,0,4,2	2,2,2,0,4,2	2,2,2,0,4,2
forced/2/1 forced/2/2 forced/2/3 forced/2/4 forced/2/5	2,2,2,0,4,2 2,3,1,0,4,2 2,3,1,0,4,2 2,2,2,0,4,2 2,3,1,0,4,2	3,2,0,2,1,1 7,3,2,4,2,1 5,3,1,2,1,1 4,2,0,3,1,1 5,3,1,2,1,1	0,0,2,2,5,2 0,- 2,3,4,6,2 0,- 1,2,2,5,2 0,0,2,3,5,2 0,- 1,2,2,5,2	6,5,3,3,6,2 9,7,6,6,9,2 6,5,3,3,6,2 5,5,2,4,6,2 6,5,3,3,6,2	 9,7,6,6,9,2
chain/2/1 chain/2/2	2,1,1,0,0,0 2,2,0,0,0,0	2,1,1,0,0,0 4,3,1,0,0,0	0,- 1,2,0,0,0 0,- 1,1,0,0,0	9,6,8,6,9,2 11,8,7,6,9,2	 11,8,8,6,9,2
open_window/2/1	2,2,0,0,0,0	5,3,4,0,8,2	0,- 4,4,0,8,2	24,13,20,12,27,2	**24,13,20,12,27,2**

Table 10A.10: Measures for Full Prolog of program State Transition building

Comments on tables 10A.7 –10A.10

Comparison of these results with those of tables 10A.1 – 10A.5 show the relative dependence of the specifications on mathematical symbols. The expansion is the greatest, even in proportion to the greater numbers involved, between tables 10A.3 and 10A.7 (the Z building) and the other pairs.

Mathematical Toolkit Measures

The Prolog code used to define the tools of Section 10.3.3 is given in code text 10A.7. Table 10A.11 shows the measures of these tools, which are needed to convert the High Level Prolog of specifications of Section 10A.2 to Full Prolog.

Specification of Mathematical Toolkit

```
cart_product(Set1,Set2,CP):-
               setof([X,Y],(member(X,Set1),member(Y,Set2)),CP).
    %Note: when using PDQ, the round brackets in cart_product/3 were
changed to square brackets.
dist_union([Set],Set).
dist_union([Set1,Set2|Sets],Union):-
    union(Set1,Set2,Union1),
    dist_union([Union1|Sets],Union).

dom(Rel,Dom):- setof1(X,Y^(is_related(Rel,X,Y)),Dom).

intersect(Set1,Set2,Intersect):-
    setof1(X,(member(X,Set1),member(Y,Set2)), Intersect).

intsfrom(F,L,F):- F=< L.
intsfrom(F,L,I):- F<L,F1 is F + 1,
    intsfrom(F1,L,I).

is_related(Rel,X,Y):- member([X,Y],Rel).

is_a_relation(Rel,A,B):-
    cart_product(A,B,CP),
    subset(Rel,CP).

isZseq(Seq):-dom(Seq,Dom),
    length(Seq,N),
    setof(Y,intsfrom(1,N,Y),Dom).

member(X, [X|Tail]).
member(X, [Y|Tail]):- member(X,Tail).
```

Code text 10A.7: The Mathematical Toolkit (contd. over)

```prolog
nthofList(I,List,Element):-
    nthofList1(0,I,List,Element).     %introduce & initialize counter

nthofList1(Counter, I, [Head|Tail],Head):-
    Counter1 is Counter + 1, %evaluate
    Counter1 = I. %test and terminate
nthofList1(Counter,I, [Head|Tail],Element):-
    Counter1 is Counter + 1,
    nthofList1(I,Counter1,Tail,Element). %continue recursion
nthofZseq(I,Seq,Element):-
    member([I,Element],Seq).

powerset(Set,Pset):- setof(Sub, subset(Sub,Set),Pset).

override(I,NewElement,List,NewList):-
    override1(0,I,List,NewElement,NewList).
                                        %introduce & initialize counter
override1(Counter,I,NewElement,[_|Tail],[NewElement|Tail]):-
    Counter1 is Counter+1,
    Counter1=I.                         % test and terminate
override1(Counter, I, NewElement, [Head|Tail],[Head|Tail1]):-
    Counter1 is Counter+1,
    override1(Counter1,I,NewElement,Tail,Tail1).

rng(Rel,Rng):- setof1(Y,X^(is_related(Rel,X,Y)),Rng).

set_difference(Set1,Set2,Diff):-
    setof1(X, (member(X,Set1),not member(X,Set2)),Diff).
subset(X,Y):- forall(member(J,X),member(J, Y)).

union(Set1,Set2,Union):-
    setof1(X, (member(X,Set1);member(X,Set2)),Union).
```

Code text 10A.7 (cont.): The Mathematical Toolkit

Clause	Ph	Pb	Pe	M1	M
cart_product/3/1	3,3,0,0,0,0	6,5,2,0,0,0	0,- 2,2,0,0,0	6,5,4,0,2,1	**6,5,4,0,2,1**
dist_union/2/1	2,1,1,0,1,1	0,0,0,0,0,0	2,1,1,0,1,1	2,1,1,0,1,1	
dist_union/2/2	2,4,0,0,1,1	5,5,1,0,1,1	0,- 1,1,0,2,1	6,4,3,0,3,1	**6,4,3,0,3,1**
dom/2/1	2,2,0,0,0,0	6,4,2,0,0,0	0,-2,2,0,0,0	5,3,3,0,2,1	**5,3,3,0,2,1**
intersect/3/1	3,3,0,0,0,0	6,5,1,0,0,0	0,- 1,1,0,0,0	6,5,3,0,2,1	**6,5,3,0,2,1**
intsfrom/3/1	3,2,1,0,0,0	2,2,0,0,0,0	0,0,1,0,0,0	2,2,1,0,0,0	
intsfrom/3/2	3,3,0,0,0,0	5,4,1,0,0,0	0,- 1,1,0,0,0	5,4,1,0,0,0	**5,4,1,0,0,0**
is_a_relation/3/1	3,3,0,0,0,0	5,4,1,0,0,0	0,- 1,1,0,0,0	10,7,8,0,4,1	**10,7,8,0,4, 1**
is_related/3/1	3,3,0,0,0,0	2,3,0,0,1,1	0,0,0,0,1,1	2,2,1,0,2,1	**2,2,1,0,2,1**
isZseq/1/1	1,1,0,0,0,0	9,4,4,1,0,0	0,-4,4,1,0,0	14,7,8,1,2,1	**14,7,8,1,2, 1**
member/2/1	2,2,1,0,1,1	0,0,0,0,0,0	2,2,1,0,1,1	2,2,1,0,1,1	
member/2/2	2,3,0,0,1,1	2,2,0,0,0,0	0,0,0,0,1,1	2,2,0,0,1,1	**2,2,1,0,1,1**
nthofList/3/1	3,3,0,0,0,0	4,3,0,1,0,0	0,0,0,1,0,0	4,4,1,1,1,1	**4,4,1,1,1,1**
nthofList1/4/1	4,4,1,0,1,1	2,2,0,0,0,0	0,0,1,0,1,1	2,2,1,0,1,1	
nthofList1/4/2	4,5,0,0,1,1	4,4,0,0,0,0	0,0,0,0,1,1	4,4,0,0,1,1	
nthofList1/4/3	4,5,0,0,1,1	4,4,0,0,0,0	0,0,0,0,1,1	4,4,0,0,1,1	
nthofList1/4/4	4,5,0,0,1,1	4,4,0,0,0,0	0,0,0,0,1,1	4,4,0,0,1,1	**4,4,1,0,1,1**
nthofZseq/3/1	3,3,0,0,0,0	2,3,0,0,1,1	0,0,0,0,1,1	2,2,1,0,2,1	**2,2,1,0,2,1**
override/4/1	4,4,0,0,0,0	5,4,0,1,0,0	0,0,0,1,0,0	5,5,2,1,2,1	**5,5,2,1,2,1**
override1/5/1	5,5,2,0,2,1	2,2,0,0,0,0	0,0,2,0,2,1	2,2,2,0,2,1	
override1/5/2	5,6,1,0,2,1	5,5,0,0,0,0	0,0,1,0,2,1	5,5,1,0,2,1	**5,5,2,0,2,1**
relations/3/1	3,3,0,0,0,0	5,4,1,0,0,0	0,- 1,1,0,0,0	12,8,9,0,4,1	**12,8,9,0,4,1**
powerset/2/1	2,2,0,0,0,0	4,3,1,0,0,0	0,- 1,1,0,0,0	6,4,4,0,2,1	**6,4,4,0,2,1**
rng/2/1	2,2,0,0,0,0	6,4,2,0,0,0	0,- 2,2,0,0,0	5,3,3,0,2,1	**5,3,3,0,2,1**
set_difference/3/1	3,3,0,0,0,0	6,4,2,0,0,0	0,- 2,2,0,0,0	6,4,4,0,2,1	**6,4,4,0,2,1**
subset/2/1	2,2,0,0,0,0	4,3,1,0,0,0	0,- 1,1,0,0,0	4,3,3,0,2,1	**4,3,3,0,2,1**
union/3/1	3,3,0,0,0,0	4,3,1,0,0,0	0,- 1,1,0,0,0	4,3,2,0,1,1	**4,3,2,0,1,1**

Table 10A.11: Data measurements for Prolog tools

Index